HISTORY OF TECHNOLOGY

HISTORY OF TECHNOLOGY

History of Technology

Volume Fifteen, 1993

Edited by
Graham Hollister-Short and
Frank A.J.L. James

MANSELL

First published 1993 by
Mansell Publishing Limited, *A Cassell imprint*
Villiers House, 41/47 Strand, London WC2N 5JE, England
387 Park Avenue South, New York, NY 10016-8810, USA

British Library Cataloguing-in-Publication Data

History of technology.—15th volume (1993)
1. Technology—History—Periodicals
609   T15

ISBN 0-7201-2160-4
ISSN 0307-5451

Library of Congress Catalog
Card Number: 76-648107

Set by Colset Private Limited, Singapore
Printed and bound in Great Britain by
Biddles Ltd, Guildford and King's Lynn

Contents

Editorial

In our editorial for Volume 14 we said that we hoped to present two papers on the subject of invention, an area of research largely ignored by historians of technology despite its obvious importance. We felt under some obligation to redeem our promise and are pleased that we have been able to do so. The present volume opens with two papers on technological invention, one of which, although modesty forbade us to say so at the time, is by one of us. Michael Lewis's wide-ranging paper, 'The Greeks and the early windmill', also addresses the subject of invention, albeit in the course of a broad approach designed to assemble materials relating to possible times and paths of diffusion from east to west of the fundamental features of the vertical windmill. What he has to tell us here reminds us strongly of what has been said earlier by other scholars about the Islamic cultural area as being a kind of magazine of technological know-how upon which medieval Europe could draw, as its own improving levels of organization and cultural sophistication made this technically and economically possible. The way in which Western scholars beat a path to Toledo after its recapture in 1085 in the course of the *Reconquista* serves as a reminder of how strong the appetite of the West then was: it would be strange to suppose that the zeal to learn was not a broad-wave phenomenon and did not wash into other domains of activity as well.

The paper on the Norsk Bergverksmuseum and the Kongsberg silver mines by Dr. Björn Berg is, we hope, the first in a sequence of papers which will together constitute a useful source of reference to some of the major archives of technological materials and collections of preserved artefacts in Europe. Now that the barriers which made travelling beyond the Iron Curtain such a toilsome and often unpleasant business have been removed, it is time that important sites such as Wieliczka and Banská Stiavnica should become more widely known. Our hope is, therefore, that the series we have in mind, beginning in Norway, will be able to include them also in due course, and thus, in a small way, assist in the reintegration of Europe. Our experience of the reopened upper levels of the mines of Kutna Hora in Bohemia and of working in the mining archives of central Slovakia in Banská Stiavnica, among the immense riches preserved there, is not one we shall soon forget.

Walter Endrei's presentation of Count Batthyány's technological excursions of the late eighteenth century on the Danubian waterways is also a reminder of this wider Europe. Batthyány's efforts are particularly interesting in that they illustrate that central eastern Europe was the scene of the very kind of activity that was taking place at much the same time in England and in the infant United States, namely, the resurrection of a late

Roman scheme for saving manpower. E.A. Thompson, in his 1952 edition of the anonymous *De rebus bellicis*, notes that at Yarmouth, in eastern England, soon after 1800, four horses in a file were used to drive the seven-foot-diameter paddle wheels of a ferry boat, propelling the craft at the rate of about 9 km per hour. It was indeed knowledge of the Yarmouth ferry which stimulated our interest when Professor Endrei first delivered his paper in Vienna in 1991.

What has also proved to be quietly diverting, while driving and trying to find a place to park in London, has been to observe oneself confirming (if that is the right word) the experience of the walking bees so nicely related in Hermann Knoflacher's paper. Altogether, we hope that the catholicity of the present offering will be to our readers' taste. Such a mix, we can guarantee, will be a feature also of at least the volume currently in preparation.

Graham Hollister-Short
Frank A.J.L. James

The Contributors

Björn Ivar Berg of the Norsk Bergverksmuseum, Postboks 18, 3601 Kongsberg, Norway, has undertaken much archaeological work in the exploration of historical mining horizons at Kongsberg.

Philippe Braunstein is Director of the École des Hautes Études en Sciences Sociales (EHESS), Boulevard Raspail, Paris, France.

W. Bernard Carlson is Associate Professor of Humanities in the School of Engineering and Applied Science at the University of Virginia. A historian of technology, he specializes in nineteenth-century American inventors and the electrical industry.

Walter Endrei is Professor at the Eötvös Loránd University, Budapest, and is head of the Centre for History of Technology of the Hungarian Academy of Sciences. His address is H-1134 Budapest, Angyalfoldi u 24/B, Hungary.

Graeme Gooday is Royal Society/British Academy Post-doctoral Research Fellow in the History of Science and Technology, Modern History Faculty, University of Oxford, Broad Street, Oxford OX1 3BD, England. He is especially interested in the historical sociology of experimental physics and electrical engineering.

Michael E. Gorman is Associate Professor of Humanities in the School of Engineering and Applied Science at the University of Virginia. A psychologist, he has published half a dozen experimental studies of scientific reasoning.

Graham Hollister-Short is Honorary Lecturer in the History of Technology, History of Science and Technology Group, Imperial College, London SW7 2AZ, England.

Alex Keller is a Senior Lecturer in the Department of History, University of Leicester, Leicester LE1 7RH, England.

Hermann Knoflacher is Professor of Traffic Planning and Traffic Engineering, Institut für Verkehrsplanung und Verkehrstechnik, Universität Wien, Karlsplatz 13, A-1040 Vienna, Austria.

Michael J.T. Lewis is a Lecturer in the Department of Adult Education, University of Hull, 60 Hardwick Street, Hull HU5 3PJ, England.

Matthew M. Mehalik is a graduate of the University of Virginia in aerospace engineering. While at Virginia he assisted Professors Carlson and Gorman with their research. He is currently employed as an engineer with Martin Marietta but he plans to attend graduate school in the history of technology in the near future.

Michael Oblon is a recent graduate of the University of Virginia in electrical engineering. In his undergraduate thesis, 'The Design of a Voice Recognition System as a Case Study for Cognitive Mapping the Inventive Process', he used Carlson and Gorman's mapping techniques to organize and guide his own efforts to design a voice recognition system. He is currently employed in the US Patent Office.

Notes for Contributors

Contributions are welcome and should be sent to the editors. They are considered on the understanding that they are previously unpublished in English and are not on offer to another journal. Papers in French and German will be considered for publication, but an English summary will be required. The editors will also consider publishing English translations of papers already published in languages other than English. Three copies should be submitted, typed in double spacing (including quotations and notes) with a margin on A4 or American Quarto paper. Include an abstract of 150–200 words and two or three sentences for 'Notes on Contributors'.

It would be appreciated if normal printers' instructions could be used. For example, words to be set in italics should be underlined and *not* put in italics. Authors who have passages originally in Cyrillic or oriental scripts should indicate the system of transliteration they have used. Quotations when long should be inset without quotation marks; when short, in single quotation marks. Spelling should follow the *Oxford English Dictionary*, and arrangement H. Hart, *Rules for Compositors* (Oxford, many editions). Be clear and consistent.

All papers should be rigorously documented, with references to primary and secondary sources typed separately from the text in double spacing and *numbered consecutively*. Cite as follows for books:

1. David Gooding, *Experiment and the Making of Meaning: Human Agency in Scientific Observation and Experiment* (Dordrecht, 1990), 54–5.

Subsequent references may be written:

3. Gooding, *op. cit.* (1), 43.

Only name the publisher for good reason. For theses, cite University Microfilm order number or at least Dissertations Abstract number. Standard works like DNB, DBB may be thus cited.

And as follows for articles:

13. Andrew Nahum, 'The Rotary Aero Engine', *Hist. Tech.*, 1986, 9: 125–66, p. 139.

Line drawings should be drawn boldly in black ink on stout white paper, feint-ruled paper or tracing paper. Photographs should be glossy prints of good contrast and well matched for tonal range. The place of an illustration should be indicated in the margin of the text where it should also be keyed in. Each illustration must be numbered and have a caption. Xerox copies may be sent when the article is first submitted for consideration.

Alexander Graham Bell, Elisha Gray and the Speaking Telegraph

A Cognitive Comparison

MICHAEL E. GORMAN, MATTHEW M. MEHALIK, W. BERNARD CARLSON AND MICHAEL OBLON

ABSTRACT

This paper begins by describing a cognitive framework for understanding the process of technological invention and then applies it to a case. On the same day that Alexander Graham Bell submitted a patent for a speaking telegraph, Elisha Gray submitted a caveat for the same sort of device. Rather than viewing this as a case of either simultaneous invention or outright theft, our framework suggests that the two inventors were following distinct problem-solving paths and viewed their final products differently. This claim is buttressed by a detailed consideration of the competition between Bell and Gray over the multiple telegraph.

INTRODUCTION

On 14 February 1876 the US Patent Office received two documents describing how the human voice could be sent and received over a telegraph line. The first, submitted by a teacher of the deaf, Alexander Graham Bell, was a formal application for an 'Improvement in Telegraphy', and it included a 'speaking telegraph' (Figure 17). A few hours later, a manufacturer of telegraph instruments, Elisha Gray, submitted a preliminary application or caveat for his speaking telegraph (Figure 22). How was it that two men from different backgrounds came to file patent documents for nearly the same thing on the same day?[1]

Yet there is still another coincidence. When Bell filed his patent application for a telephone he had only a conception of it, not a working device. Consequently, after filing his patent in Washington Bell returned to Boston and began a series of important experiments.[2] In the course of these

1

experiments, Bell succeeded in the first transmission of human speech on 9 March 1876. Significantly, success came not with the device Bell described in his application but with a liquid transmitter similar to Gray's apparatus (Figure 23). Did Bell somehow learn about Gray's device?

What is one to make of these coincidences? Should one simply marvel at this classic case of simultaneous invention and forgo any analysis? Should one admit that the constraints of nature and business often narrow the range of technological choices so that there is only 'one best way', and in this case two men happened upon the same solution? Or should one suspect that either Bell or Gray stole ideas from the other? Since that fateful day in February 1876 the inventors, their lawyers and historians have investigated and debated the possibility of wrongdoing.[3] Because the telephone quickly came to be the basis of a highly profitable business, lawyers for both Bell and Gray spent nearly twenty years trying to establish legal priority to this invention in US courts.[4]

It is thus very tempting to reduce this case to one of two alternatives: either Bell and Gray were constrained by external factors which led to simultaneous inventions or one man stole the speaking telegraph from the other. However, in this paper we wish to challenge these two alternatives and propose a new way of looking at this case. What if Bell and Gray were following their own distinct lines of thought toward separate goals? Is it not possible that each man had a different conceptualization of a device involving sound and electricity and it is only in hindsight that we conclude that they both invented the telephone?

To investigate Bell's and Gray's individual lines of thought means attempting to reconstruct the mental and hands-on activities which constituted the invention process for each of them. In undertaking this cognitive reconstruction we are departing from the dominant mode of investigation in the history and sociology of technology today which downplays the inner, creative processes of invention and discovery and instead emphasizes the processes by which individuals and society establish the legitimacy of technological inventions or scientific claims.[5] This emphasis on the social context of technological change and social construction has greatly enriched the history of technology in the last two decades by permitting the field to move beyond naive technological determinism and by providing a common set of issues for scholars to debate.[6] Yet, taken to its extreme, an exclusive focus on social factors has resulted at times in technological determinism being replaced by social determinism in which historical actors are seen as simply the carriers of interests, values, or goals.[7] This is especially true in the case of Bell and Gray, where David Hounshell has explained their differences in terms of the social roles played by each man and not by examining the content of their ideas and actions.[8] What has been lost sight of is the fact that individuals think and act independently and unpredictably, sometimes in support of their social context and sometimes in opposition.

We believe that the analysis of invention as a social process must be complemented by an investigation of invention as a cognitive process. Obviously, an essential task for the history and sociology of technology is to understand the interplay of social context and technological content. To do so requires a thoroughgoing investigation of technological content; we must understand what ideas and objects individuals bring together in creating new machines and techniques. To know both the ideas and objects and the ways in which individuals join them, we must come to terms with how they represent ideas and objects in their minds as well as on paper and as artefacts.[9] These representations may be generated by the individual or taken from his or her society and culture. In our view, there should be no conflict between cognitive and social approaches to technology; both are needed if we are to understand in a substantial way how individuals and groups shape technology and society.[10]

In this essay, we will develop a cognitive framework for invention by investigating how Bell and Gray developed speaking telegraphs in early 1876. We will look closely at their work in multiple telegraphy during the previous seven years, analysing the ideas, objects and practices they acquired. With this background established, we will then examine how each man conceptualized a speaking telegraph and prepared patent documents. To complete our comparative analysis, we will discuss briefly what Bell and Gray did after February 1876 to realize their inventions.

In narrating the rivalry between Bell and Gray, we find it is more important to focus on the differences in their cognitive processes than on the similarity of the artefacts they created at one point in time. Indeed, differences in process give us clues to differences in the artefacts; as we shall see, Bell and Gray did not invent the same device. By applying a cognitive framework to this episode in technological history we hope to show that a cognitive analysis is a necessary complement to understanding the social context of technology.

A COGNITIVE FRAMEWORK FOR UNDERSTANDING INVENTION

Our cognitive framework[11] has four major components:

- *Mental models* A mental model is a dynamic, visual representation of a potential device that an inventor can 'run' in his or her mind. For example, Edison began his kinetoscope, or motion picture, invention with a mental model of a machine that would do 'for the Eye what the phonograph does for the Ear'.[12] Like the phonograph, Edison intended that the kinetoscope would be used by individuals for both recording and viewing moving pictures. Consequently, this mental model led Edison to develop a peep show and not a projecting machine. There is a growing literature on mental models in cognitive psychology, but this work does not focus on inventors.[13]
- *Slots* We divide an inventor's mental model into areas of concentration which we call 'slots', following the cognitive psychology literature.[14] Slots are our reconstruction of the way an inventor subdivides a problem. With the kinetoscope, Edison's mental model suggested that he needed a component analogous to the sound groove on the recording cylinder on his phonograph. Consequently, Edison initially sought a way to create a spiral of tiny photographs on a revolving drum.

• *Mechanical representations* Inventors have familiar devices or 'stock solutions' that they insert in slots in their mental models.[15] A slot is analogous to a variable; a mechanical representation is analogous to a value for a variable. For example, Edison's phonograph mental model did not include a mechanism for interrupting motion so that the eye could register the illusion of a smooth flow of images. He therefore needed to create a slot for a device that would cause the cylinder to pause briefly. In this slot, he inserted a double-action pawl he had used on his stock tickers to convert circular motion into linear motion; we refer to this double-action pawl as a mechanical representation.[16] Because an inventor may manipulate these devices not only on the workbench but also in his or her imagination or in sketches, we refer to these devices as representations and not simply as components or objects.

• *Heuristics* Inventors also employ different problem-solving strategies.[17] Edison, for example, often delegated different parts of a project to his assistants, based on his skills and theirs. On the kinetoscope, he assigned William K.L. Dickson the task of perfecting the photographic elements of the kinetoscope, while Edison himself worked on the electromechanical parts.[18]

At this point, our framework is like a mental model early in the invention process: it is incomplete, unstable and vague. Nevertheless it is useful because it alerts us to look for certain aspects of the invention process and allows us to compare inventors using a common language and set of concepts. We can thereby gain a better understanding of similarities and differences in cognitive style among inventors. For example, the framework helps us understand the relative roles of Edison and Dickson on the kinetoscope. Edison provided the overall mental model, the major research heuristic and nearly all of the mechanical representations. Dickson did valuable photographic work, but this does not justify the claim of historians that he invented the kinetoscope.[19] Nor does it permit us to label Edison the sole inventor. Instead, the framework redirects our attention from 'who gets the credit?' to 'what did each participant do?' and 'how did he represent what he was doing?'.

To organize and depict Bell and Gray's paths to the speaking telegraph, we have created flowcharts or multi-level branching tree diagrams that include every sketch or artefact we can locate. We refer to these flowcharts as maps because they reflect our reconstruction of each inventor's path to an eventual goal that neither set out deliberately to find. These maps are implemented on a Macintosh computer, therefore they are highly flexible and can be continually revised in the light of new information. These maps were used as a kind of 'database' in creating this narrative, and portions of these maps will be used as illustrations in this paper.[20]

BELL'S MENTAL MODEL FOR A MULTIPLE TELEGRAPH

The history of the telephone began not with a desire to transmit speech but with the economic need to send multiple messages over a single telegraph wire. During the middle decades of the nineteenth century, inventors and businessmen perfected the electric telegraph and established it as a major form of rapid communication. By the mid-1870s, this technology was largely controlled by Western Union, one of the first corporate monopolies in the United States. Yet, in building a nationwide telegraph network Western Union was hampered by a severe problem: as the volume of messages grew so the cost and complexity of the network grew even more quickly. In response, the telegraph giant encouraged investors to develop a variety of new devices, including schemes whereby several messages could be sent simultaneously over a single wire. In 1872 Western Union adopted Joseph Stearns's duplex (two-message) system, and it was soon clear that fame and fortune awaited the inventor of a four- or eight-message system.[21]

Among the several American inventors in the 1870s who tried to develop a multiple telegraph was Alexander Graham Bell. In pursuing this invention, Bell was actively encouraged by his future father-in-law, Gardiner Hubbard.[22] Hubbard was bitterly opposed to Western Union, which he viewed as a monopolistic giant. Hubbard hoped to slay the giant by having Bell develop a multiple-message system which could be used to create an alternative telegraph network.

Bell pursued a unique path to a multiple telegraph because his understanding of electricity differed from that of his competitors. Others who worked on multiple telegraphy, including Gray, were experts in the field of telegraphy. Bell, however, had no formal education or experience with electricity. Instead, he had an extensive background in phonetics, elocution and musical theory. Consequently, Bell's multiple telegraph was based upon his knowledge of musical theory, which proved to be an invaluable tool, aiding his inventive process. In fact, his first insight into a possible multiple telegraph system grew out of his understanding of how a piano operated; as Bell explained in 1876:

> If we press down the pedal of a piano, and sing into the instrument, the sound waves cause that string to resound which corresponds to the note sung. Now, conceive that under each string is placed an electromagnet, and that all the magnets are united in one circuit. If we transmit a series of electrical pulses along the circuit (corresponding in number and regularity to the vibrations of a sound), a similar series of attractive impulses will appear at the magnets, and the piano-string whose rate of vibration corresponds will resound . . .
>
> [If] many notes are sung simultaneously into the piano, each system of sound waves affects its corresponding string as readily as though the other systems had no existence: hence, however many different series of attractive impulses are made to appear simultaneously at the magnets, each one must affect its corresponding string as though it

came alone. The study of sympathetic vibrations thus led me to the conclusion that a large number of telegraphic messages could be sent along the same circuit, without confusion, if the signals for each message had a certain definite pitch different from those employed for the others . . . [23]

This observation provided Bell with an initial mental model for a multiple telegraph; it was a device he could not build, but he could imagine how it might work. In order to translate this mental model into a working device, Bell needed to understand how sound could be converted into electrical signals, and then back again to sound. Bell tried very hard to be scientific in his approach.[24] At one point, he attempted to build a device that would artificially produce vowel sounds, only to find that he was replicating the work of Hermann von Helmholtz.

Bell learned about the acoustical research of Helmholtz from Alexander J. Ellis, an English phonetician who was a friend of Bell's father.[25] Ellis explained to Bell that Helmholtz had constructed a device which artificially produced vowel sounds. By carefully listening to several singers utter vowels and comparing these pitches to a response on tuning-forks, Helmholtz found that the voice was composed of harmonic partials. These experiments showed him that the human voice was actually composed of several resonant pitches. These tones therefore could be simulated by a series of vibrating tuning-forks corresponding to the same tonal frequencies as the vowel sounds. Bell's diagram of the Helmholtz apparatus is shown in Figure 1. Bell described this device as follows:

Helmholtz had demonstrated the compound nature of the vowel sounds by producing them artificially by a synthetical process. For example, he would cause the simultaneous vibration of three tuning-forks of different pitches—one of these would represent the pitch of the voice—and this fork he caused to vibrate in front of a resonator tuned to its own pitch, so as to cause it to produce a loud musical tone. The other two forks corresponded in pitch to the front and back cavities of the mouth in uttering some vowel sound. These forks were caused to resound very faintly. The simultaneous vibration of the three forks produced one loud, fundamental sound, and two higher partial tones. The effect upon the ear was as though someone sang a vowel sound. In an interview with Mr. Ellis, he attempted to describe to me the apparatus used by Helmholtz. Helmholtz kept his forks in vibration by means of electro-magnets and a voltaic battery; but I found that I had not sufficient electrical knowledge to understand the arrangement used by Helmholtz. I therefore determined to study electricity . . . and study Helmholtz's researches and repeat his experiments.[26]

Bell transformed Helmholtz's apparatus into a mental model for a multiple telegraph by viewing the lower interrupting tuning-fork as a transmitter, and the upper resonating fork as a receiver. If a telegraph key were inserted between the interrupting and resonating forks, one could alternately send and interrupt the musical tones. In effect, this new mental

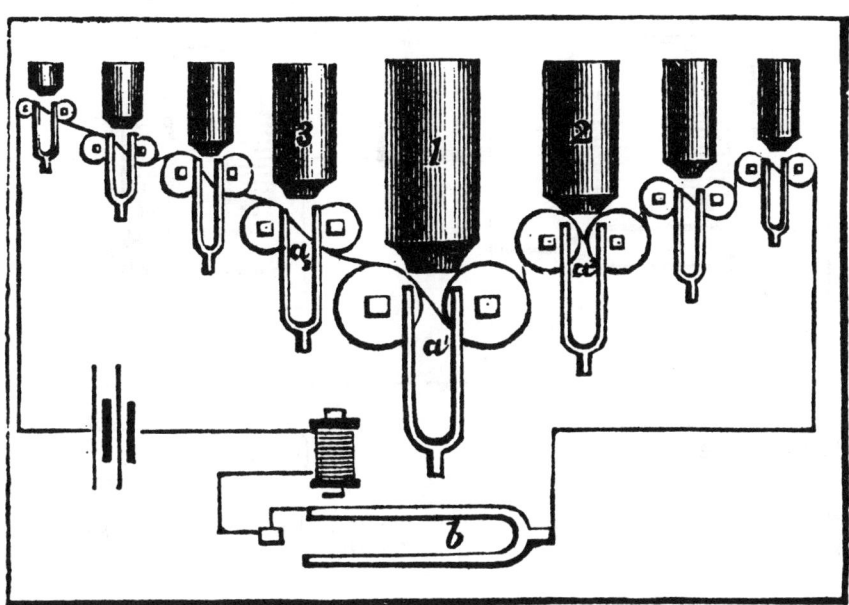

Figure 1 Bell's representation of the Helmholtz apparatus. The transmitter consisted of tuning-fork b with a battery and an electromagnet above its upper prong. The receiver consisted of a series of tuning-forks a, each with its own electromagnet and resonant cavity. When the tuning-fork b vibrated, a needle attached to its upper prong alternately made and broke contact with a cup of mercury, causing current to flow through the electromagnet and attract the upper prong. When the needle broke contact with the mercury, the current was interrupted, the magnetic attraction stopped, and the prong dropped back into the mercury. In this way, the continuous vibrations of tuning-fork b were converted into an intermittent current. In the receiver, the tuning-forks were kept in constant motion, vibrating in frequency corresponding to both the size of the fork and the period of the intermittent current. By placing resonant cavities above each tuning-fork the vibrations became audible, and by adjusting the aperture of the chamber the volume could be adjusted. If the forks were tuned properly, they would respond both to the primary tone of tuning-fork b and its overtones. By partially opening and closing the proper combination of chambers on the top, Helmholtz could produce different harmonic structures and simulate vowel sounds. Source: G. B. Prescott, *Bell's Electric Speaking Telephone* (New York, D. Appleton & Co., 1884), p. 66.

model represented an alternative to his original one, in which each tuning-fork would play a role similar to that of a string on the piano.

Sometime in 1873 Bell built the telegraph circuit shown in Figure 2.[27] In this experiment, Bell reduced the Helmholtz apparatus to one transmitting and one receiving fork; if he got this circuit to work, he believed he could add more transmitters and receivers and create a multiple telegraph. For his transmitter Bell arranged a tuning-fork, an electromagnet and a mercury cup in the same fashion as the lower fork in Helmholtz's device. As the current flowed from the battery through the electromagnet,

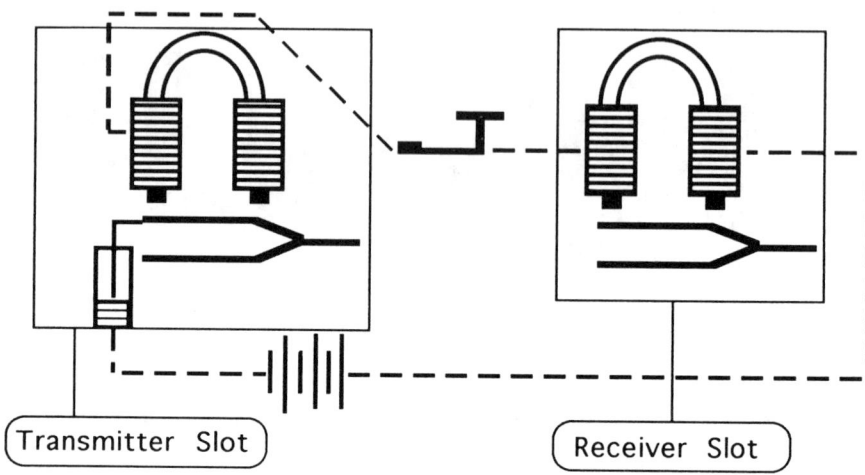

Transmitter Slot   Receiver Slot

Figure 2 Slots in Bell's mental model of a multiple telegraph. Underlying drawing is adapted from Alexander Graham Bell, *The Multiple Telegraph* (Boston: Franklin Press, Rand, Avery & Co., 1876), Figure 2.

the tuning-fork was attracted upward, which interrupted the circuit by raising a needle connected to an upper tine of the fork out of the mercury cup. When the current stopped, the electromagnet was de-energized, the fork was no longer attracted upward, and the needle made contact again with the mercury. This would re-establish the circuit, and the process would then be repeated. With this arrangement, Bell secured an intermittent current at the frequency of the tuning-fork.

Bell's receiver in this experiment worked in a similar fashion. As the current flowed through the circuit, the electromagnet on the right side was energized, attracting the tuning-fork and causing it to vibrate. By placing a resonating cavity (not shown) near the receiver's tuning-fork, Bell heard the receiving fork respond to the transmitting fork. By connecting a telegraph key between the devices, Bell could control the current flowing between the transmitter and receiver, create dots and dashes, and hence send messages.[28]

Superimposed on Bell's sketch are two boxes that group the device into its major slots, or functional categories. Some slots are common to all inventors working on a particular problem. For example, all multiple telegraph inventors experimented with devices designed to function as transmitters and receivers; therefore we can organize their efforts into transmitter and receiver slots. This arrangement serves as a useful tool for comparison. As we shall see, Bell and Gray each had different mental models for what might go in these slots and different mechanical representations to try in the slots.

For Bell, the Helmholtz interrupting fork and upper resonating fork served as mental models for what ought to go in the transmitter and receiver slots. Yet try as he might, Bell was unable to get the apparatus in Figure 2 to work very well. 'On account of the small size of the tuning-forks, and the imperfect means at my command,' he reported, 'the vibrations obtained were not as satisfactory as had been hoped.'[29] Consequently, Bell embarked on a long series of experiments to fill these slots in a way consistent with his mental model. In doing so he developed new mechanical representations which could be substituted into the slots. For example, Bell substituted a vibrating reed from a bassoon in place of the tuning-fork in the transmitter slot, a platinum contact fixed to the end of the reed dipped in and out of a dish of mercury. Testing this device, he could see that the reed afforded advantages over the tuning-fork in terms of adjustability and sensitivity. After reading about an acoustic telegraph in a popular book on electricity, Bell made another substitution and replaced the tuning-forks with thin steel reeds.[30]

The upper sketch in Figure 3 shows how Bell simply substituted steel reeds for the tuning-forks in both his transmitter and receiver. After making this substitution, Bell then modified his transmitter and receiver by removing the resonant boxes and by inverting the reeds and electromagnets. He also removed the mercury cup from the transmitter and replaced it with a thumbscrew which made or broke contact with the vibrating steel reed (see the lower sketch in Figure 3). By adding this thumbscrew to his receiver, Bell made his transmitter and receiver identical. In doing so, Bell merged the transmitter and receiver slots into a kind of 'transceiver' slot. This reed relay could be substituted in both the transmitter and receiver slots. This transceiver slot was unique to Bell for, as we shall see, Gray used distinct devices for transmitters and receivers in most of his experiments.

Although Bell found that this new reed relay responded better than his tuning-fork design, he continued to experiment with other devices suited to separate transmitter and receiver slots. For example, in October or November 1874, he inserted in the transmitter slot a series of magnets arranged around a cylinder which induced a rapid intermittent current when rotated.[31]

In these experiments, Bell focused increasingly on how an electric current could be converted to sound and vice versa. For instance, in November 1874, while using the rotating magnet transmitter, he noticed that an intermittent current produced a sound in the iron core of the electromagnet in his receiver. He attributed this phenomenon to molecular vibration: when magnetized, the molecules in the metal core moved closer together then spread farther apart. Bell conducted further experiments with this phenomenon, substituting a helix for the iron core of the electromagnet and placing nails and other objects in the middle of the helix to enhance the effect.[32]

This device was similar to a receiver which Philipp Reis had invented in Germany in the 1860s and which he called a 'telephone'.[33] Bell claimed that he learned about Reis's device only after he had completed these experiments.[34] Bell found that when an iron nail was placed between two cylindrical pieces of iron surrounded by a helix of wire 'a note of similar pitch to that produced by the transmitting instrument proceeded from R [the device].' The instrument, however, was not sensitive to one pitch

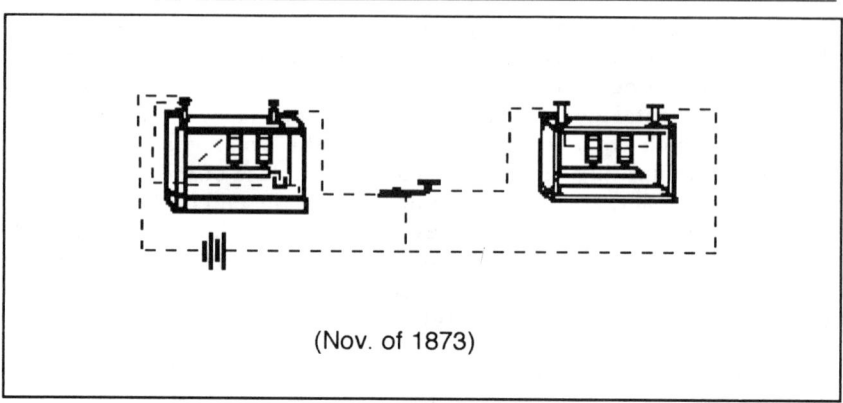

(Nov. of 1873)

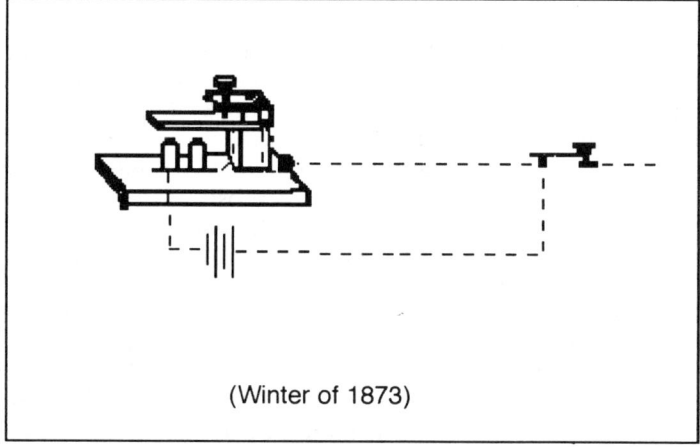

(Winter of 1873)

Figure 3 The evolution of Bell's reed relay, 1873. The upper drawing shows how Bell first substituted steel plates for the tuning-forks in his transmitters and receivers. Note that the mercury cup was still present in the transmitter. The lower drawing shows a steel reed suspended over the electromagnets without the sounding box or the mercury cup. It was this device, typically with a single electromagnet, that became one of Bell's most useful mechanical representations. Redrawn from Bell, *op. cit.* (Figure 2), Figures 5 and 7.

alone, but reproduced very loudly the 'unison of whatever transmitter was employed.'[35] Bell had developed a universal receiver.

At this point, Bell had several mechanical representations with which to experiment: his reed relay, rotating magnet transmitter and a universal receiver. With these devices in hand, he turned now to investigating different circuit arrangements. These arrangements involved a variety of slots between transmitter and receiver, including batteries, line resistance, and whether the connections between multiple devices are in series or in parallel.[36] As we shall see, Bell struggled to connect several transmitters and receivers in a workable fashion and, in this struggle, he altered his mental model in significant ways.

Bell initially used his reed relay transceiver to send two tones across a telegraph line (Figure 4). In these experiments, one tone was received clearly but the second was not. Bell reasoned that the steel reeds in the receivers were not properly tuned. This failure prompted him to make an important decision about his style of invention: 'It became evident to me, that with my own rude workmanship, and with the limited time and means at my disposal, I could not hope to construct any better models. I therefore from this time (November, 1873) devoted less time to practical experiment than to the theoretical development of the details of the invention.'[37]

Although Bell did not succeed in this initial multiple-telegraph experiment, he did notice a phenomenon that helped him understand how the vibrations of the reeds or plates corresponded with the electrical pulses and musical tones. When Bell set the transmitters into vibration, he pressed his ear against each receiver to see if there were any corresponding sympathetic vibrations. He heard 'two musical tones, corresponding in pitch to the two transmitters employed, but different in pitch from the sound produced when the reed of the receiver at [the] ear was plucked with the finger'.[38] This showed Bell that the vibrating devices could emit tones other than their natural vibrating frequencies. Although Bell did not use this information immediately, this discovery, along with his experiments with his universal receiver, was another clue toward understanding how a single device could respond to multiple pitches.

In his first multiple telegraph designs, Bell could only send messages in one direction (i.e. from Station A to Station N in Figure 4, but not from N to A). Bell knew that a successful multiple telegraph would have to send messages in both directions. To achieve bidirectional communication, Bell wanted to use his reed relays as transceivers. To accomplish this, he decided to have his reed transceivers induce a current in the primary circuit of an induction coil; when a telegraph key was closed, this circuit would be connected with the main line via a secondary circuit, allowing the impulses to be transmitted.[39] After assembling a device based on this principle in December 1873, Bell ran into another obstacle. He found that this circuit arrangement created twice as many electrical impulses in the secondary circuit as in the primary. Furthermore, these impulses were of opposite polarity, which had the effect of reversing the polarity of the receiving instruments at each impulse. The solution, Bell decided, was permanently to magnetize his steel reeds: 'In order to render the reeds of the receiving instrument permanently magnetized, I proposed to attach them to the poles of a permanent magnet, and allow the free ends of the reeds to be attracted and repelled by the poles of an electro-magnet.' Bell further saw that

> when a permanent magnet is moved towards the pole of an electro-magnet, a current of electricity appeared in the coil of the electro-magnet; and that when the permanent magnet was moved from the electro-magnet, a current of opposite kind was induced in the coils.

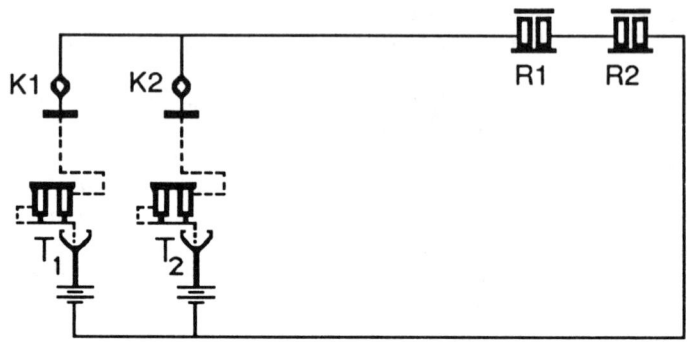

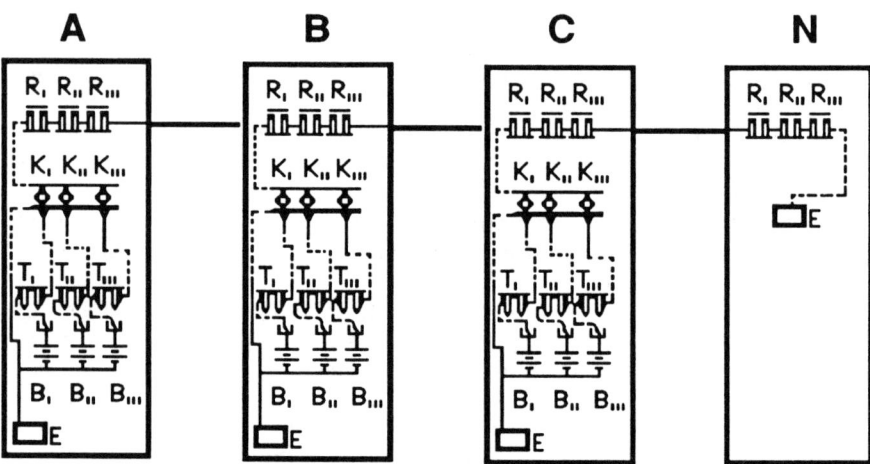

Figure 4 Bell's plan for a multiple-harmonic telegraph. The upper drawing shows Bell's basic circuit for sending two tones over the same wire. T1 and T2 were reed-relay transmitters which could be placed in the circuit by telegraph keys K1 and K2. As K1 for transmitter T1 was pressed, the periodic intermittent current in the local circuit (transmitter and battery) flowed on to the external line. Because receiver R1 was tuned to the same frequency T1, it was supposed to respond only to T1's tone. In the lower drawing, Bell shows how this basic circuit could be used in a number of telegraph stations (A through N) to send three simultaneous messages. Each station would have three transmitters (T_i, T_{ii}, T_{iii}), three telegraph keys (K_i, K_{ii}, K_{iii}), three receivers (R_i, R_{ii}, R_{iii}), and three batteries (B_i, B_{ii}, B_{iii}). A single telegraph line joined all of the stations, and each station had a ground return (E). When the key controlling any of the transmitters was pressed, only the receivers with the same corresponding frequency were supposed to respond. Using this arrangement, Bell was able to send three messages in one direction, from A to N. Redrawn from Bell, *op. cit.* (Figure 2), Figures 6 and 9.

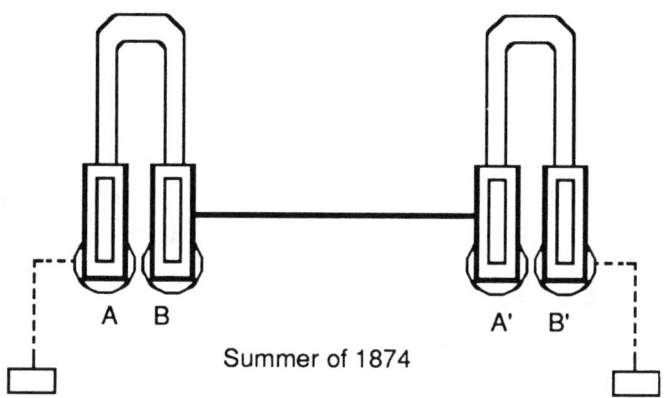

Figure 5 Bell's plan for a multiple-harmonic telegraph using permanent horseshoe magnets. A, B, A′, and B′ were steel reeds attached to the poles of two horseshoe magnets. Underneath each of the reeds were electromagnets. As each reed vibrated in front of the electromagnet, it induced an undulating current in the circuit. A and A′ corresponded in pitch, as did B and B′. When A was set into vibration, A′ was set in corresponding vibration, reflecting both the pitch and amplitude of the musical sound produced by A, while B and B′ remained silent. In the same fashion, B set B′ into vibration without affecting A and A′. Also, a vibration in A′ induced a corresponding vibration in A, so that the communication was bi-directional. Redrawn from Bell, *op. cit.* (Figure 2), Figure 15.

> I had no doubt, therefore, that a permanent magnet, like the reed of one of my receiving instruments, vibrating with the frequency of a musical sound in front of the pole of an electro-magnet, should induce in the coils of the latter alternately positive and negative impulses corresponding in frequency to the vibration of the reed, and that these reversed impulses would come at equal distances apart.[40]

Figure 5 illustrates Bell's idea of how his steel reeds might be joined to the poles of a permanent magnet while vibrating over the poles of an electromagnet. Significantly, this device utilized what Bell called an undulating current, not an intermittent current. The current would have the form of a sinusoidal wave. From Helmholtz, Bell knew that this curve 'express[ed] in a graphical manner the vibratory movement to the air while the reeds were producing their electrical tones'.[41] Furthermore, the vibrations of individual reed transceivers could be summed into a single undulating curve. The necessary circuit connections, however, would be extremely complex and beyond Bell's skill. Therefore he simplified them by imagining what would happen if all the reeds were placed over a single electromagnet.

In the summer of 1874, Bell sketched a device consisting of a series of steel reeds over a single electromagnet. The principle is the same as in Figure 5, but it differed in that Bell replaced the single reed over each pole of the electromagnet with a large number of reeds. Like the strings of a piano, these reeds would reproduce musical tones. When one spoke a vowel

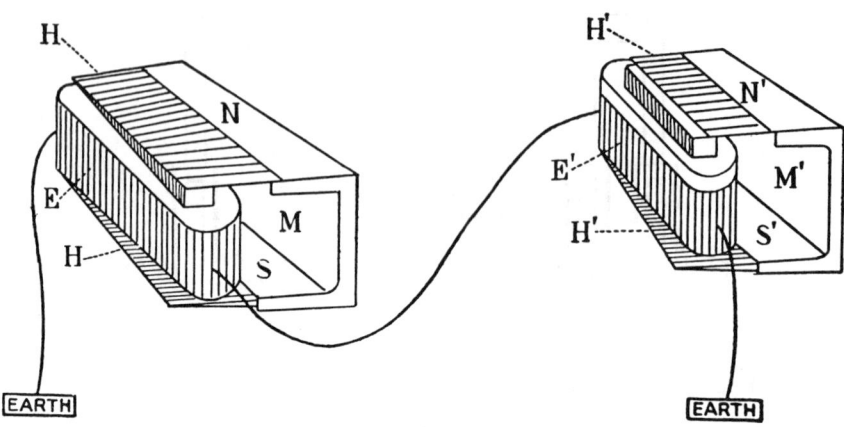

Figure 6 Bell's harp apparatus. M and M′ are permanent horseshoe magnets, to which a long series of steel reeds (H and H′) were attached to each pole. E and E′ were electromagnets connected by a wire. Bell theorized that if one sang or spoke in front of one set of reeds, various reeds would vibrate and induce a current in electromagnet E. This current was then transmitted over the wire, caused electromagnet E′ to attract and repel the same reeds along H′, and thus reproduce the original sound. *Source*: Frederick Leland Rhodes, *Beginnings of Telephony* (Harper & Brothers, 1929), p. 11.

into the transmitting harp, Bell visualized how a combination of reeds representing the fundamental tone and its overtones would vibrate and this exact combination would be transmitted to the other side, reproducing the vowel sound. This principle had been clearly established by the Helmholtz device which was Bell's original mental model; in this case, however, the single interrupting fork and series of separate resonators were replaced by a series of reeds combining to induce a current in a single electromagnet.

Bell knew he could never build such a device, owing in part to the multiplicity of reeds that would be required, but it served as a new mental model for a universal transceiver—this harp apparatus could transmit and receive speech, musical tones, or any other pattern of sounds.

Bell encounters Elisha Gray

A few months after he conceptualized his harp apparatus, Bell learned that he was not alone in devising a multiple telegraph using acoustical tones. A manufacturer of telegraph instruments from Chicago, Elisha Gray, was pursuing a similar line of research. As Bell recalled:

> It was just at this time [Sept. or Oct. of 1874] . . . that I first heard of Mr. Elisha Gray. I received a note from my friend Dr. Clarence J. Blake, in which he alluded to a letter received from Mr. Elisha Gray, descriptive of experiments made with Prof. Tyndall relative to the telegraphic transmission of vocal sounds. Dr. Blake stated his desire to show me the letter, and expressed his wish to have me meet

Mr. Gray. I called upon Dr. Blake next day, and told him that, as
I was at that time applying for a caveat for an invention which would
ultimately lead to the telephonic transmission of vocal sounds, I
thought it might be well for me to be ignorant of Mr. Gray's researches
until I had secured my patents. For this reason, I did not see the letter;
and I am still ignorant of its contents. A day or two after seeing Dr.
Blake, I again heard of Mr. Gray through my solicitor, Mr. Adams.
He informed me that Mr. Gray was applying for patents upon a
method of transmitting sound telegraphically, and that he had had a
conversation with Mr. Gray's solicitor, Mr. Hayes, relative to our
several inventions. I do not know what passed at this conversation; but
though I had no reason to suppose that the confidence between counsel
and client was violated, I believe that Mr. Adams had unintentionally,
by the mention of my invention, given Mr. Hayes a hint which at once
set Mr. Gray upon my track.

So far as I have found out from Mr. Gray's patents, his invention
at that time (end of September or beginning of October, 1874) consisted of nothing more than a method of transmitting sound through
living tissue, no claim being laid to the practical application of
telephony to the simultaneous transmission of messages along a single
wire.[42]

ELISHA GRAY'S PATH TO A MULTIPLE-HARMONIC TELEGRAPH

Bell was indeed correct in his observation that Gray had been concentrating
on investigating the transmission of sounds through animal tissue. A director and electrician of the Western Electric Manufacturing Company, Gray
had invented several telegraph devices including a repeater, relay and printing telegraph.[43] With backing from Western Electric and several successful inventions under his belt, Gray was a serious threat to Bell in the race
for a multiple telegraph and, later, a speaking telegraph.

Gray was especially interested in how electric currents changed as
they passed through the human body or other animal tissue. In 1874
he experimented with how animal tissue, while being rubbed on a conducting surface, audibly reproduced frequencies sent by a transmitter. Figure
7 shows Gray's basic apparatus and the slots where he tried different
mechanical representations. Whereas Bell merged the transmitter and
receiver slots into a transceiver slot, Gray kept them separate in his multiple
telegraph experiments. Moreover, the animal tissue slot was unique to
Gray.

In his first animal tissue experiments in early 1874, Gray substituted
different mechanical representations into the transmitter and receiver slots
(Figure 8). In particular, he placed a vibrating electrotome in the transmitter slot and a zinc-lined bathtub in the receiver slot.[44] Gray grabbed
one lead from the vibrating electrotome and placed his other hand on
the surface of the conducting bathtub. Gray rubbed his hand at different
speeds and with different pressures across the surface of the bathtub. The
bathtub resounded with the same pitch which was produced by the coil and

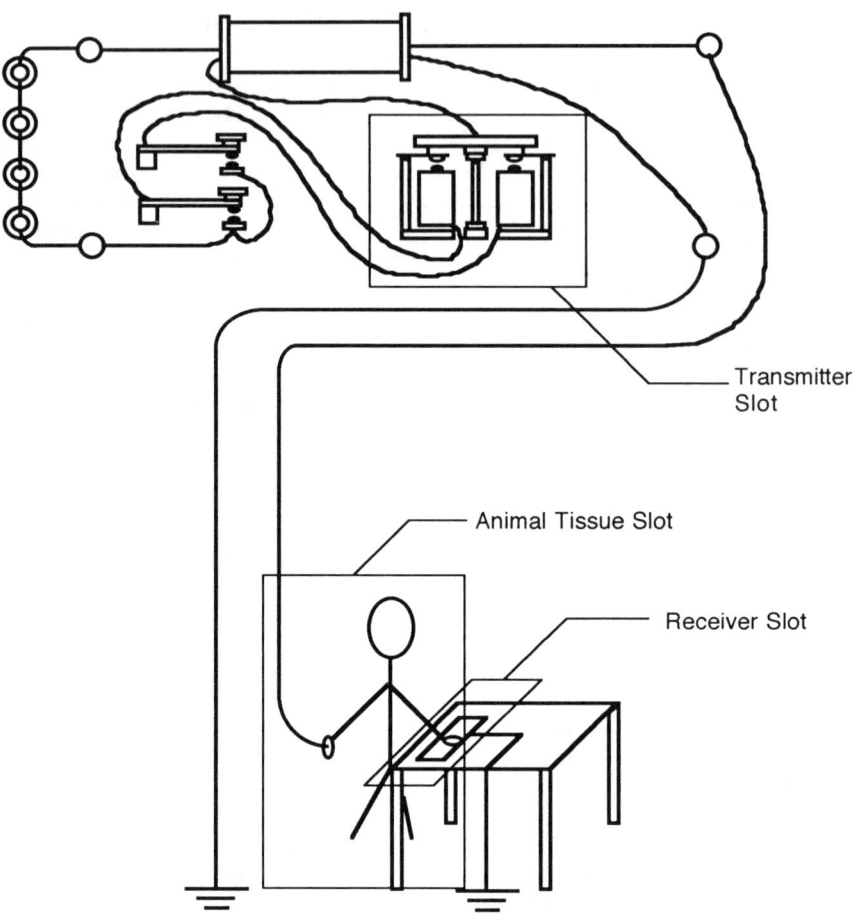

Figure 7 Slot diagram of Gray's animal tissue experiments. Redrawn from Elisha Gray, 'Electric Telegraph for Transmitting Musical Tones', US Patent 166,096 (filed 19 January 1875, granted 27 July 1875).

traversed Gray's body. Significantly, Gray concluded from this experiment that he was able to vary the quality and the pitch of the tone resonating from the bathtub.[45]

In April or May 1874 Gray conducted a second experiment by inserting two different devices in his slots. In the transmitter slot he placed a two-tone transmitter that was based on his mental model which combined the vibrating electrotome and Reis transmitter.[46] Gray desired to send an undulating current over a telegraph line, and the two-tone transmitter accomplished this by having two electrotome-like devices. The transmitter consisted of two single-pole electromagnets, each with a vibrating armature (Figure 9). Each armature made and broke contact with a platinum point

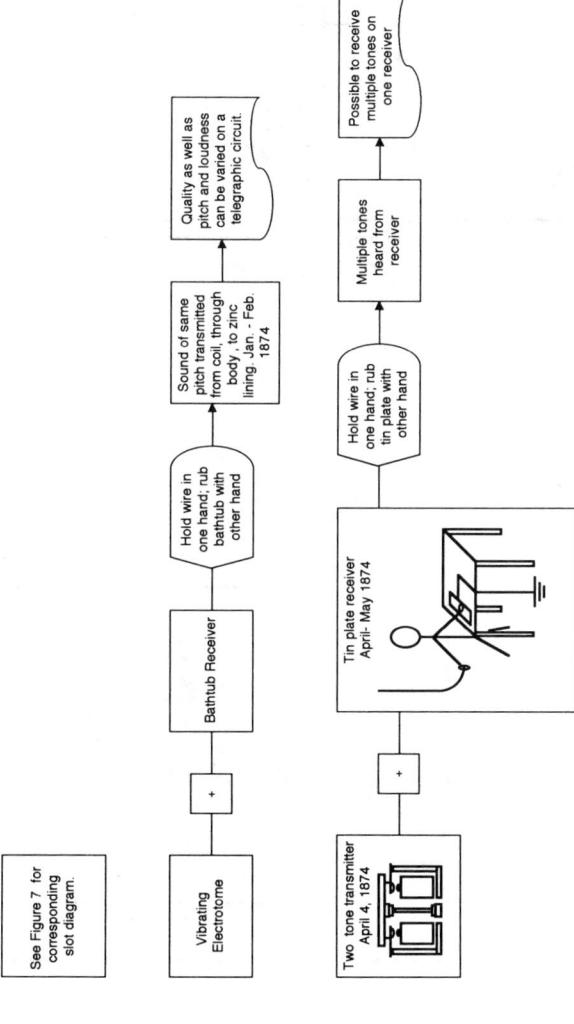

Figure 8 Gray's animal tissue experiments, 1874. Redrawn from Gray patent, *op. cit.* (Figure 7).

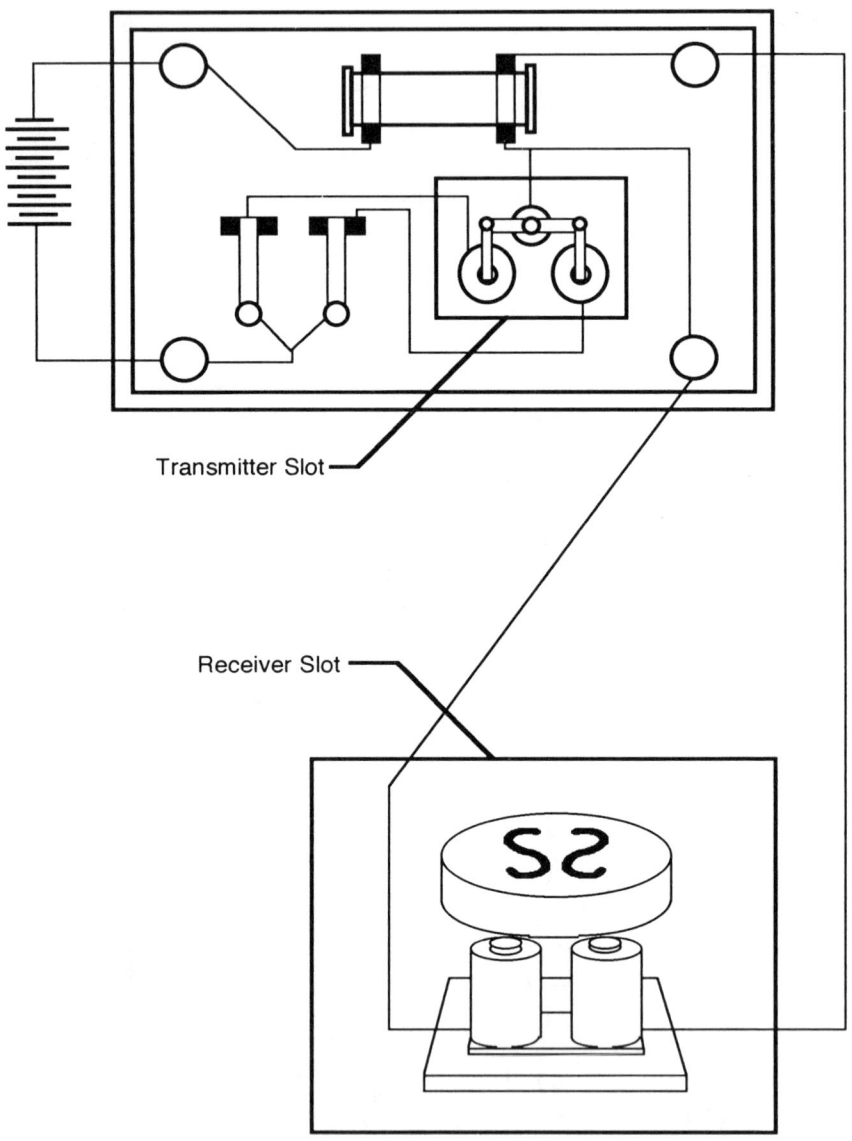

Figure 9 Gray's two-tone transmitter and tin-box receiver. Redrawn from Gray Patent 166,095.

which switched the current to the coil on and off. Because each electromagnet had a different electrical resistance, each exerted a different magnetic pull on its armature and thus caused each armature to vibrate at a different frequency. Each coil and armature combination was controlled by its own telegraph key, so that each frequency could be sent separately or simultaneously. To boost the strength of the current, these electromagnets were connected to an induction coil. This induction coil functioned like a modern transformer and stepped up the current before it was sent out on to the telegraph line.

In his second animal tissue experiment Gray used a receiver which consisted of a grounded piece of galvanized tin. Gray again placed himself in the animal tissue slot. During this experiment the two-tone transmitter sent two different, audible tones over a telegraph line, which were reproduced on the tin-plate receiver as Gray rubbed his hand on the plate. Gray also tried a variety of animal tissues, ranging from oyster shell to leather.[47] As we shall see, trying different combinations of mechanical representations was a heuristic Gray frequently employed.

From his animal tissue experiments, Gray realized that it was possible to send multiple tones over a single wire. Using this insight, Gray began working on a multiple-harmonic telegraph in which each message was transmitted by a different tone.[48] To realize this invention, Gray conducted numerous experiments between 1874 and 1876, and these experiments are summarized in Figures 9 and 10. Figure 9 shows the slots in which Gray tried a multitude of transmitters and receivers. A discussion of each of these experiments would take many pages, so we will discuss only the most important.

The first experiments further refined Gray's mental model of how tones and pitches could be sent over a telegraphic circuit. In this series, shown in Figure 10, Gray inserted a two-octave transmitter in the transmitter slot.[49] This device could send twenty-four different pitches (two octaves) over one telegraphic circuit. Each tone was generated by a single-tone transmitter. The single-tone transmitter worked like one of the two electromagnets from the two-tone transmitter. Each single-tone transmitter was tuned to a different pitch. Gray often used several single-tone transmitters inside more complex devices capable of sending multiple tones, such as his two-octave transmitter and printing telegraph. Because he used the single-tone transmitter by inserting it into slots in different inventions, it became one of Gray's mechanical representations.[50]

Figure 10 shows that Gray also tried several receivers in combination with his two-octave transmitter. Gray claimed that his methodology was 'systematic', and we are calling this his combination heuristic:

> Having all these uses in my mind, and supposing I had secured in my first patent [filed 27 June 1874] all the various applications that might be made in the matter of transmitting sounds telegraphically, I pursued my investigations in a systematic way, placing each development to the credit of the *particular application* to which it seemed to belong.[51]

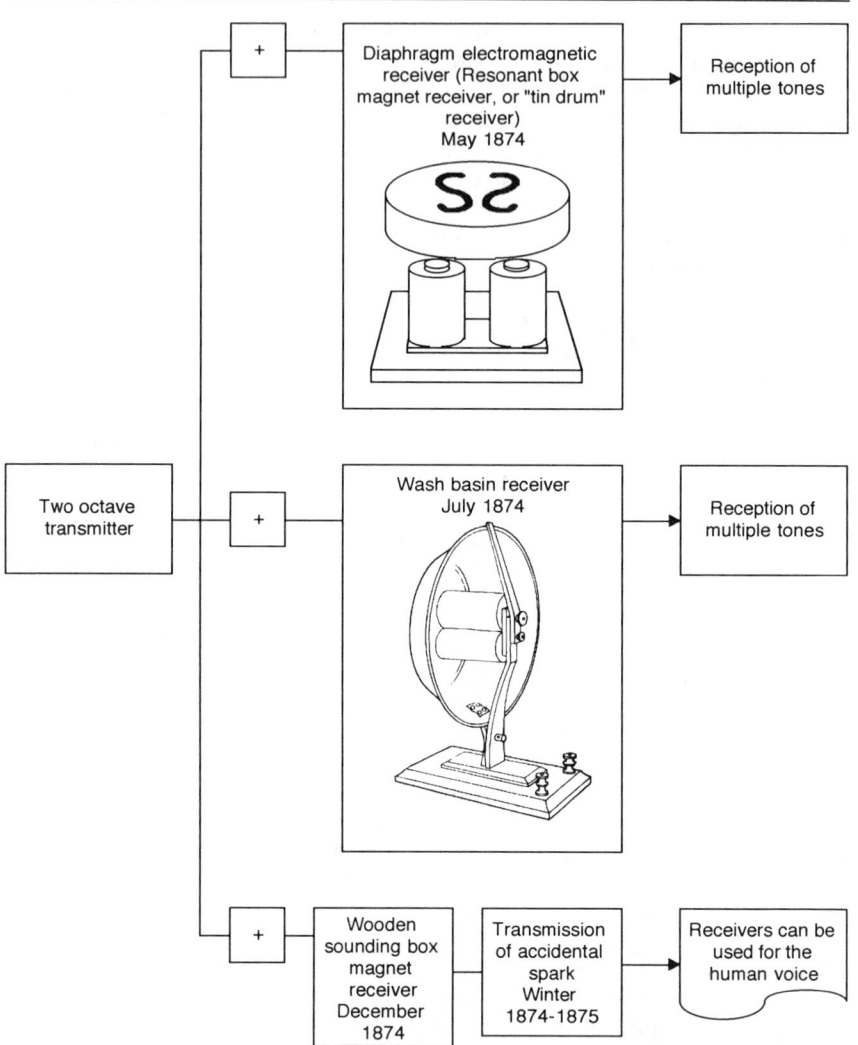

Figure 10 Different receivers used by Gray in combination with the two-octave transmitter. Tin receiver figure is redrawn from Gray's US Patent 166,095. The illustration of the wash-basin receiver is from David A. Hounshell, 'Two Paths to the Telephone', *Scientific American*, January 1981, 244: 156–63, p. 159.

So Gray tried his various receivers in combination with his two-octave transmitter. All of these receivers were based on his mental model of the Reis receiver. According to Gray, the principle of the Reis receiver was that 'when a coil of wire surrounding a bar of iron or the core of an electromagnet is traversed by an electric current, the said bar will be slightly elongated, and if these currents succeed each other with sufficient rapidity, a vibratory motion will be given to said bar, and it will give forth a musical tone'.[52] All of Gray's receivers embodied this principle and hence were capable of reproducing several tones simultaneously, but they employed different mechanical representations in the amplification of the vibrating core of the electromagnets. So, while the Reis receiver functioned as Gray's receiver mental model, these mechanical representations came from several other sources. The polar relay furnished the mechanical representation of the double-pole electromagnet found in receivers Gray developed during the first half of 1874. The resonant cavities in his receivers came from various sources. The tin drum came from a magnetic adaptation of the resonant cavity of a violin.[53] The wash-basin receiver was perhaps based on his previous experience with using a bathtub as a receiver.

Gray conducted the series of experiments shown in Figure 10 in order to test the two-octave transmitter and to explore the possibilities of sending multiple tones over a circuit. In each experiment Gray was able to send and receive multiple tones. In the course of these experiments Gray evolved a mental model for how his multiple telegraph (and later his telephone) would use multiple transmitters and tones. Notice that Gray used transmitters capable of sending multiple tones over a telegraph wire. These transmitters, however, employed several sub-component transmitters, each of which generated only one tone. Gray's multiple telegraph was based on these individual transmitters, and he patented a system of sending multiple messages over a wire in one direction at a time. Each message was sent from a separate transmitter at a different frequency. Significantly, this mode of transmission was different from the way a telephone works: typically, voice transmission occurs by using a transmitter which reproduces complex sounds and not from several transmitters each producing a single pitch.

During one experiment in the winter of 1874–75, an accidental spark was generated in the two-octave transmitter (Figure 10). Gray claimed that this spark produced a sound similar to the human voice, and it was transmitted and received over the telegraph circuit. Gray now realized that it might be possible to send and receive complex sounds, such as the human voice, over a telegraph line.[54]

Gray explored this insight by devising a new transmitter not based on his Reis transmitter/vibrating electrotome mental model. This new device was a mechanical transmitter, and it permitted Gray to investigate the possibilities of transmitting speech across a telegraph line. We shall discuss these experiments and the corresponding changes in Gray's mental model in a moment, but before doing so, we wish to compare Bell and Gray's multiple telegraphs and discuss whether they were simultaneous inventions or stolen ideas.

BELL AND GRAY'S HARMONIC MULTIPLE TELEGRAPHS: SIMULTANEOUS INVENTION OR STOLEN IDEA?

In October 1874, when Bell claimed he first heard of Gray's work, he had just sent Gardiner Hubbard a patent caveat for a multiple-harmonic telegraph.[55] This document described 'the application of acoustical principles to telegraphy so as to permit of the simultaneous transmission of a large number of messages along a single wire without confusion with one another'. The caveat covered most of Bell's devices discussed above, but Bell especially wanted to claim the general principle of using 'armatures having definite rates of vibration as a means of distinguishing between signals sent separately or simultaneously along the same circuit'.[56] As part of the caveat, Bell swore that he believed himself to be the original inventor of this mode of transmission.

After preparing this caveat, Bell read an article by Hayes, Gray's patent attorney, which alluded to 'the simultaneous transmission of messages by means of musical notes, as an invention recently made by Mr. Gray'.[57] The article went on to describe Gray's apparatus, which Bell thought seemed highly similar to his own. Shocked, Bell wondered if his own lawyer, Adams, might have given Hayes 'a hint which at once set Mr. Gray upon my track'.[58] Let us try to reconstruct the state of affairs in order to evaluate whether these two men were working on the same kind of device and, if so, whether Gray needed to get any ideas from Bell.

On 10 July 1874 an article appeared in the *New York Times* reporting Gray's successful demonstration of the transmission and reception of musical tones across a telegraph line. With his two-octave transmitter and various receivers, Gray had transmitted and received familiar tunes such as 'Yankee Doodle' over a telegraph line of 2400 miles. The article highlighted Gray's ability to transmit notes and chords over long distances, which was something Bell had never attempted to do. In closing, the article suggested that these instruments could be used for telegraphy. Gray called this invention a telephone, and the article claimed that he had filed for patents in the United States and Europe. This article was subsequently reprinted in the *Boston Commonwealth* on 14 November 1874, where Bell read it.

What is not clear from the article is whether Gray had the capability to transmit, receive and decode simultaneous telegraph *messages*. Sending and receiving multiple tones is one thing, but multiple messages is an entirely different matter. Was the invention described in the article a multiple-harmonic telegraph? Had Gray devised a way to analyse or decode messages sent using simultaneous carrier tones?

The same question might be asked of Bell. Hubbard told him not to file the caveat:

> The papers say Mr. Gray has obtained Patents for the transmission of musical sounds and etc. I have sent to Washington a request to have them sent to you. A patent is of no value provided another made the invention at an earlier date and the issue of a patent to A does not prevent B from subsequently obtaining a patent for the same invention if he was the prior inventor.[59]

Hubbard made it clear that a caveat would guard only against those who came after Bell; it would provide no protection against an invention filed previously.[60]

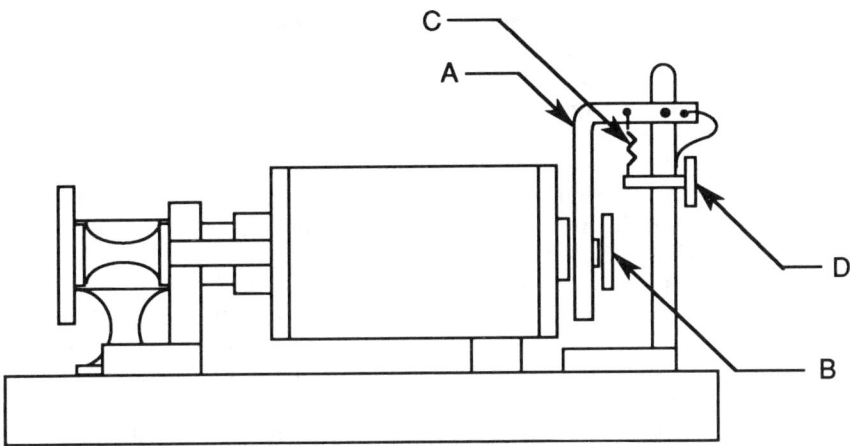

Figure 11 Cross-sectional side view of Gray's analysing receiver, 1875. (See Figure 12 for top and front views of this device). The device consisted of an electromagnet, a long, tuned metal reed B, and a light-weight, elbow-shaped contact A. C was a small coiled spring for adjusting A. When the electromagnet received pulses of current at a specific frequency, B would vibrate at that frequency. As B vibrated, A rattled because it could not keep up with B's motion. As a result, A would make-and-break contact with B, and hence convert the continuous vibrations of B into an intermittent current. This intermittent current in turn was sent to a Morse sounder, which produced dots and dashes for the operator to hear. By using several of these analyser receivers (each with B tuned to a different frequency), Gray was able to send and decode several telegraph messages over a single line. Redrawn from E. Gray, 'Improvement in Receivers for Electro-Harmonic Telegraphs,' US Patent 166,094 (filed 28 June 1875, granted 27 July 1875).

Bell eventually submitted a patent application for his multiple telegraph on 25 February 1875, two days after Gray submitted one for his.[61] These closely timed submissions foreshadow the competition over the telephone, in which Gray filed a caveat on the same day Bell filed a patent. The records of the earlier controversy are skimpier, but the patents awarded show how close the two inventors were, and allow us to discuss their simultaneous inventions in terms of our framework.

These February applications reveal that by 1874 both Bell and Gray could send musical tones over a wire, and that Gray could do so with greater proficiency than Bell. However, before this accomplishment could be used for multiple-message telegraphy, the tones had to be decoded into distinct signals. What was needed was a device that would transform tones into individual messages. Consequently, we will next discuss how Gray developed such a device and then compare it to Bell's efforts.

Gray's analysing receiver
Figure 11 shows Gray's analysing receiver from his February 1875 application, the application he submitted two days before Bell. The major feature of this device was the combination of an elbow-shaped lever and regulator

spring.[62] The use of a lightweight contact lever with a slower rate of vibration resting against a tuned spring was Gray's solution to the problem of how to send multiple simultaneous messages over a single wire but decode them into separate signals for messages. As will be shown, Bell also employed a lightweight circuit-breaker lever in his patent application filed two days after Gray. In order to settle the question of whether or not Gray independently developed this receiver, we must look at the origins of Gray's mental model and the mechanical representations that Gray used in developing this analysing receiver.

Recall that as early as April 1874 Gray had succeeded in transmitting two tones using his two-tone transmitter and tin-drum receiver. Gray built upon this initial success by developing his two-octave transmitter and a family of receiving instruments capable of receiving the tunes that Gray would play over long-distance telegraph lines.[63] With these instruments Gray gave several demonstrations in New York and Washington in May and June 1874, after which he returned home to Chicago.

Gray claimed that, upon returning to Chicago, he worked on the problem of being able to sort out individual messages sent simultaneously over a telegraph line. The record of Gray's efforts between the middle of June and 24 July, when he left Chicago for a trip to Europe, is incomplete.[64] Gray later claimed that during this period he conceived a device consisting of a tuning-fork and a double-pole electromagnet. The handle of the tuning-fork was mounted on one pole of an electromagnet, while the tines projected over the other pole. Gray used this device in connection with his multiple-tone transmitter, presumably his two-octave transmitter, and 'this was a step in the direction of effecting an analysis of composite tones when transmitted through the wire'.[65]

The principle Gray was after in this device could have come from a study of Helmholtz. But Gray, who is typically explicit about his sources, such as his knowledge of the Reis telephone, never mentioned Helmholtz. Gray might have obtained this information from Bell, through the Adams–Hayes connection. Bell had an intimate knowledge of the Helmholtz apparatus and he frequently employed tuning-forks in his multiple telegraph experiments. However, we have no way of proving or disproving this hypothesis.

What we need to do is uncover in more detail Gray's mental model of a harmonic multiple telegraph between May 1874, when Gray performed his public demonstrations, and February 1875, when he submitted his patent application for the analysing receiver. We also need to trace the origins of the mechanical representations used in the analysing receiver, such as a tuned bar and an elbow-shaped contact lever.

Gray's patent applications, dated 18 April 1874 and 27 June 1874, covering his animal tissue experiments and his two-tone transmitter suggest that he did not have a mental model of a receiver designed to decode tones into messages. Instead, Gray proposed that several telegraph operators would each listen to a different message being sent at a particular tone. Moreover, Gray's statement that he constructed a tuning-fork device in June or July 1874 suggests that he was working on a mechanical representation to sort complex messages automatically. Gray prepared a caveat to be executed on 6 August 1874, a day or two before his departure for Europe, but he was later vague about its contents:

> This caveat was hastily prepared just on the eve of departure, and very imperfectly described the progress I had made at that time in transmitting several messages simultaneously on the same wire. This caveat formed the basis of a subsequent application, filed February 23, 1875; also, the one filed January 27, 1876, for a system of multiple telegraphy, based upon the ability to transmit a number of tones simultaneously over the same wire and analyze them at the receiving end, so that each tone would be audible on a particular instrument which was tuned to it, but no other.[66]

The caveat was not filed until 2 November 1874, after Gray had returned from Europe. Notice that Gray stated that the caveat 'formed the basis' for the future patent applications, but he did not mention whether the caveat discussed how to analyse transmitted signals mechanically. He claimed only the ability to transmit, not receive, several messages on the same wire.

Gray did, however, possess an analysing receiver by late November or early December of 1874. William Goodridge, Gray's assistant, recalled that by then Gray had constructed the device shown in his February 1875 patent.[67] Although Goodridge's description does not give us a clear picture of Gray's mental model for this receiver, he did leave us one important clue. Responding to concerns as to whether or not the instruments were based on the same principles for both transmitting and receiving, Goodridge explained, 'The instruments were alike in form at the two ends, but differently connected, and were not interchangeable.'[68]

This clue offers insight into Gray's mental model of the analysing receiver. On closer inspection, we see that the geometry and mechanics of the tuned analysing receiver were nearly identical to those of his vibrating reed transmitters. In fact, Gray used some of the same mechanical representations in both his single-tone transmitter and his analysing receiver. These similarities explain Goodridge's account that the devices functioned similarly but were not interchangeable. Of course this is in direct contrast to Bell, whose reed relay served as a transceiver.

Figure 12 shows Gray's single-tone transmitter and a top view of his analysing receiver, the same device depicted in Figure 11. Both instruments employed electromagnets with fluctuating currents, although the analyser used two electromagnets and the transmitter only one. Both had a tuned reed suspended in front of the poles of the electromagnets. The reed in the analyser was held at both ends with the electromagnets in the centre, whereas the transmitter reed was suspended at one end while the other end vibrated above the pole of the electromagnet. The analysing receiver used a bent contact lever with a contact point, which made and broke a circuit between the vibrating reed and lever. The transmitter used a similar object to make and break contact with the vibrating reed, a device which Gray called an 'intermeditate spring'. This device was used instead of a rigid contact point for the transmitter. The effect of the intermediate spring was that it offered 'a resistance so slight

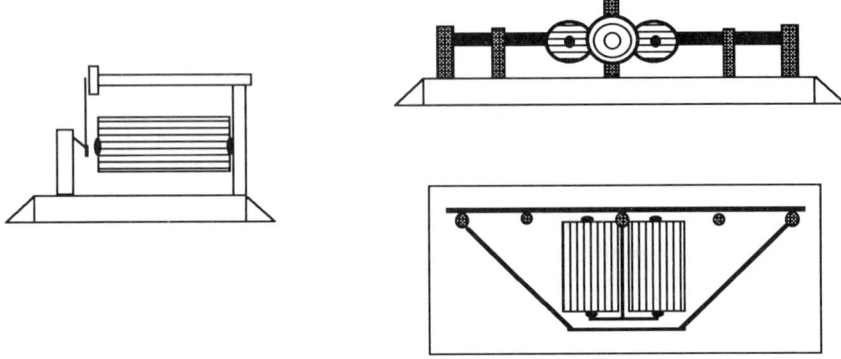

Figure 12 Gray's single-tone transmitter (left) and analysing receiver (right). The single-tone transmitter used a make-and-break contact between a vertical reed and contact point to create periodic fluctuations in the current sent onto the telegraph line. Gray called the contact point an 'intermediate spring', and he used this type of contact on all his transmitters after May 1874. The analysing receiver (see Figure 11 for a cross-sectional side view of this device) employed a long horizontal reed fixed at both ends. Bell thought that Gray's decision to fasten down both ends of the reed in his receivers was a significant difference between their receivers; see Figure 14 for Bell's sketch of Gray's receiver. Redrawn from Elisha Gray, *Experimental Researches in Electroharmonic Telegraphy and Telephone: 1867–1878*. (New York: Russell Brothers, 1878; reprinted in George Shiers, ed., *The Telephone: An Historical Anthology*, Arno, 1977), 24–5.

that it does not practically interfere with the movements of the reed.'[69] Thus, the analysing receiver included several of the same mechanical representations we have seen in Gray's transmitters developed as early as April of 1874.

The earliest description of Gray's analysing receiver, Goodridge's account, dates this device from November 1874, but it appears that Gray had all of the mechanical representations to build the analysing receiver as early as May 1874, when he had developed his single-tone transmitter. Why was there this gap of seven months? Gray needed to develop an adequate mental model for an analysing receiver. In his *Experimental Researches in Telephony*, Gray recounted an instance where he noticed that his wash-basin receiver would respond to a particular tone with 'unusual power', and he found that the tone that made the receiver respond strongly was the 'pitch of the cavity of the pan'. Gray implied that he conducted these experiments during the summer of 1874.[70] This wash-basin experiment may have given Gray a mental model of a tuned receiver which could discriminate individual tones.

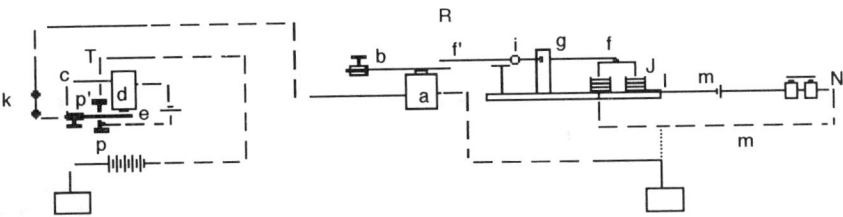

Figure 13 Bell's harmonic-multiple-telegraph circuit, March, 1875. The vibratory circuit-breaker consisted of lever f'f on pivot g with prong J dipping into mercury cups l. As reed e vibrated in the transmitter on the left, reed b responded in the receiver on the right. As this reed b vibrated, it caused lever f'f to lift, dipping prongs J into mercury cups l, which completed the circuit through the Morse sounder N, causing it to click. Thus this device converted an undulatory current into an intermittent current. Redrawn from A.G. Bell, 'Improvement in Transmitters and Receivers for Electric Telegraphs', US Patent 161,739 (filed 6 March 1875, granted 6 April 1875).

Bell's vibratory circuit-breaker

Bell tried to solve the problem of how to translate tones into telegraph signals with different mechanical representations. For example, he attached silk threads to one of his universal receivers, each thread making contact with a single tuning-fork which was tuned to correspond to a transmitting fork.[71] He also considered the possibility of using Helmholtz resonators.

Figure 13 depicts the vibratory circuit-breaker from a patent application Bell filed on 6 March 1875 for an autograph or facsimile telegraph. Bell attached this mechanical representation to his autograph telegraph partly to avoid having the Patent Office place his application into interference with Gray's invention.[72] Bell added this vibratory circuit-breaker to the receiver in his autograph telegraph in order to convert continuous vibrations from a reed into electrical pulses. This mechanical representation accomplished this by having the vibratory reed tip a lever into twin cups of mercury, completing the circuit and thereby converting a tone into a make-or-break impulse suitable for telegraphy. This lever and twin cups resemble Samuel F.B. Morse's original telegraph portrule, illustrating the way in which mechanical representations can be borrowed from another device.[73] Although the Morse instrument may have provided him with a mechanical representation that he used in his 6 March application, Bell nonetheless made it clear that it was the principle he was after, not just a specific mechanical representation:

> Many forms of circuit breakers for the purpose may be employed such as membranes &c., all that is required being that the circuit breaker shall be capable of vibratory or oscillatory movement, and that its normal rate of movement, when in oscillation or vibration, shall be slower than that of the receiver by which it is actuated.[74]

Indeed, Bell later patented an improvement in this vibratory circuit-breaker.[75]

Figure 14 Bell's sketch of the spring used in Gray's analysing receiver. Redrawn from A.G. Bell to G. Hubbard, 8 May 1875, Box 80, Bell Family Papers, Library of Congress, Washington, DC.

In this later patent for a vibratory circuit-breaker, Bell showed how this receiver could be activated by pulses from an autograph telegraph.[76] Bell also made the general claim that the same mechanisms could be used for a multiple telegraph. As in his early Helmholtz-inspired experiments, there would be a series of reed receivers, each tuned to a particular frequency. Bell pointed out that one advantage of these devices was that they allowed an undulatory current to be converted to an intermittent one for telegraph purposes.[77]

Bell studied Gray's use of vibrating reeds and springs and found there were significant differences between their devices. In a May 1875 letter, Bell inferred that Gray clamped the spring (or reed) in his analysing receiver at both ends and therefore that it responded only to notes whose ratios correspond to 1, 2, 3, and so on. In contrast, Bell's springs, clamped only at one end, 'answer[ed] only to notes whose ratios of vibration are as 1, 3, 5, 7'. Bell added that Gray's spring would require constant adjustment of the tension, whereas 'A spring such as I use has an invariable pitch dependent upon its elasticity. It cannot get out of order unless the intermittent attraction should in process of time affect its elasticity.' Furthermore, Bell noted that the pitch of a note being transmitted via intermittent current can be detected 'by placing the ear against one of the poles—and the defective armature can then be tuned to correspond'.[78] Whether Bell's account of the relative advantages of his mechanical representation over Gray's is correct or not, his discussion reveals that the two devices operated differently and therefore one inventor's device could not have been 'stolen' from the other.

What Gray might have obtained indirectly from Bell was a mental model for developing a multiple telegraph. In his articles on Gray, David Hounshell claims that the idea of using musical tones to send telegraphic messages is implicit in Gray's early devices, but we think this statement is a reflection of our ignorance.[79] What is missing is a detailed account of Gray's cognitive process at this point. How exactly did he decide to go from a device to send musical tones to a multiple telegraph? Specifically, how did Gray devise a receiver capable of responding to only one tone? Our framework suggests that Gray could have gone from his mental model for a universal receiver, based on Reis, to the idea of separate tuned receivers via his observation that the wash-basin responded better to certain pitches than others. Once he had the idea that individual tones could be received by separate devices, it is not hard to imagine him adapting his sophisticated single-tone transmitters to serve as receivers. Thus, it is not necessary to assume that Gray had to get a mental model from Bell.

The competition over the multiple-harmonic telegraph demonstrates that both Bell and Gray were working towards similar solutions to the problem of decoding multiple messages, but used different processes, and arrived at distinct devices. Therefore their inventions are 'simultaneous' only in hindsight; at the time, as Bell's May 1875 letter to Hubbard illustrates, Bell thought there were important differences between their inventions. What is possible is that each obtained important ideas from what he knew about the other's devices, although whatever information was passed on by attorneys and newspapers was inaccurate in important respects.[80]

The analysis in this section has prepared us for considering the more famous case of the speaking telegraph, to which we now turn our attention.

The ear phonautograph
Bell lacked Gray's telegraph expertise. Indeed, Bell lamented his lack of electrical knowledge in an interview with Joseph Henry at the Smithsonian Institution.[81] He did, however, possess a unique area of expertise. He was a teacher of the deaf and therefore understood the importance of speech in communication. His father, Alexander Melville, had invented a system of 'Visible Speech' intended to help the deaf learn to speak. Bell was similarly interested in devices that would help the deaf 'see' speech.

Figure 15 shows one of Bell's devices for visualizing speech, an ear phonautograph, which he built in 1874 following a suggestion from Clarence Blake.[82] When one spoke into the cone, the eardrum and stapes, malleus and incus (all taken from a preserved human ear) were set into vibration; these vibrations were traced on smoked glass by a bristle brush attached to the end of the incus. From the phonautograph, Bell gained a tactile, 'hands-on' understanding of how speech was translated into an undulating wave by the vibrations of the bones of the ear. From his multiple telegraph experiments, Bell gained a similar understanding of how the vibrations of a reed or a combination of reeds could be translated into an undulating electric current.

On 2 June 1875, Bell and his assistant, Watson, made a serendipitous discovery that suggested how these phonautograph and multiple telegraph experiences could be combined into a new mental model for a device that would transmit speech. Bell set up three multiple telegraph stations, A, B and C, each with three tuned-reed relays (see Figure 4). He wanted to be able to pluck the first reed in station A and have the corresponding reeds in stations B and C vibrate. These reeds were very difficult to tune and required constant adjustment by shortening or lengthening. Bell was supervising stations A and B, and Watson was in another room working with C. When Bell depressed the telegraph key for one of the reeds at A, the corresponding reed at B vibrated well, but Watson said C was stuck. To release it, Watson plucked it, and Bell noticed that this caused the corresponding reed at B to vibrate powerfully. Bell then listened to each of the reeds at station B in succession, placing his ear right against them,

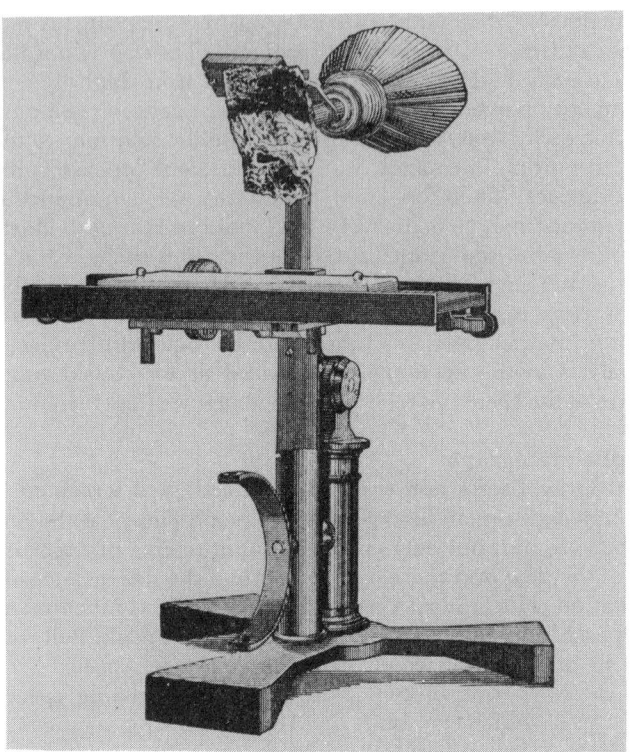

Figure 15 Bell's ear phonautograph. The device consists of a human ear mounted on a wooden frame. One spoke into the ear, causing the bones of the inner ear to vibrate; a bristle brush, connected to the bones, traced the shape of the sound waves on a piece of smoked glass, which slid slowly underneath. From Prescott, *op. cit.* (Figure 1), 69.

and heard both the pitch and the overtones of the tuned reed. Bell quickly concluded:

> These experiments at once removed the doubt that had been in my mind since the summer of 1874, that magneto-electric currents generated by the vibration of an armature in front of an electro-magnet would be too feeble to produce audible effects that could be practically utilized for the purposes of multiple telegraphy and of speech-transmission.[83]

Bell immediately asked Watson to build a working telephone in which a reed relay was attached to a diaphragm or membrane with a speaking cavity over it. As one spoke into the cavity, the membrane vibrated; these vibrations were translated into an electrical current by the reed relay sent to a similar device on the other end. Unfortunately, this device did not produce intelligible speech, though Bell and Watson heard a kind of mumbling that suggested they were on the right track. Later, in Bell's

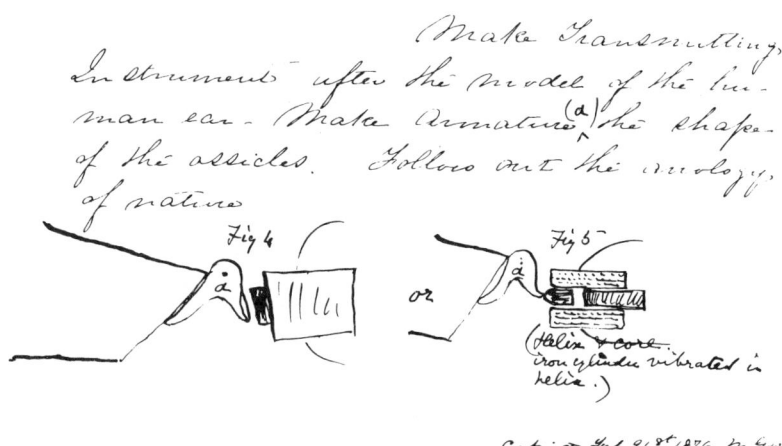

Figure 16 Bell's sketch of his ear mental model, February 1876. In both figures, Bell drew a small metal armature a which imitated the general shape of the bones of the inner ear. In Figure 4, he located this armature between a cone and diaphragm and an electromagnet. In Figure 5, Bell replaced the electromagnet with a 'Helix & core iron cylinder vibrated in helix'. From 'Experiments Made by Alexander Graham Bell (vol. 1),' notebook, Box 258, Bell Family Papers, p. 13.

notebook, we discover that the mental model guiding the construction of this device was the human ear.

The ear mental model
The first clear statement of Bell's mental model for a speaking telegraph came almost a year later in his notebook, where he sketched an ear with two different mechanical representations next to the bones of the inner ear[84] (see Figure 16). On the left is an electromagnet, suggesting that the armature and steel reed on his familiar mechanical representation (see Figure 13) will serve a function similar to the bones of the inner ear. On the right an iron cylinder is attached to the bones and this vibrates in the centre of a magnetized helix with an iron core. Bell had conducted experiments with helices and cores, verifying that such devices could produce an undulatory current.[85] Beside the sketch, Bell wrote, 'Make Transmitting Instrument after the model of the human ear. Make Armature (a) the shape of the ossicles. Follow out the analogy of nature.'[86]

This statement that Bell was 'following the analogy of nature' illustrates another component of our framework. Inventors and scientists often employ heuristics, or 'rules of thumb', to reach their goals.[87] Bell's plan of 'following the analogy of nature' is one such strategy: when in doubt, try to copy nature. Note that heuristics depend heavily on mental models and mechanical representations; in order for Bell to copy nature, he had to have both a clear understanding of the ear which came from the phonautograph, and a set of mechanical representations that he could use to transform his mental model into a working device.

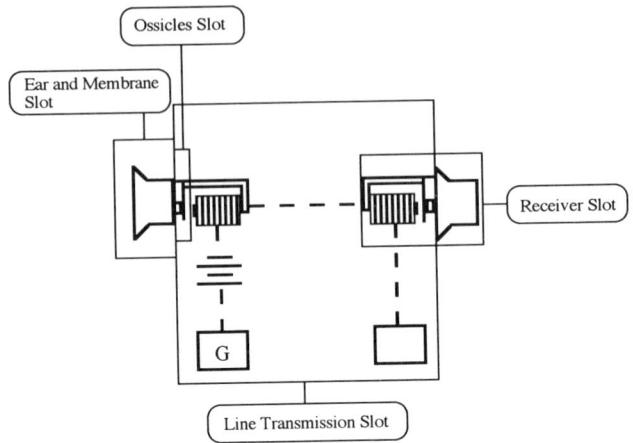

Figure 17 Slots in Bell's ear mental model. Underlying drawing is adapted from A.G. Bell, 'Improvements in Telegraphy', US Patent 174,465 (filed 14 February 1876, granted 7 March 1876).

Figure 17 shows a slot diagram drawn over one of the figures from Bell's February 1876 telephone patent. This patent became the focus of extensive litigation, and Elisha Gray was one of the leading challengers. Indeed, Gray's caveat for a speaking telegraph arrived in the patent office a few hours after Bell's patent. In subsequent sections, we will discuss both the Bell patent and the Gray caveat in detail. For now, we wish to show how our slot diagram helps organize Bell's experimental activities after his patent. Bell opened each of these slots, sometimes singly, sometimes in combination, and inserted different mechanical representations to improve performance.

Shortly after his patent, he began working in what we have labelled a 'long-distance line transmission slot'. To perfect an armature and coil that would transmit a powerful and clear signal, Bell returned to experiments with tuning-forks right after he filed his patent. He knew that if he could get improved transmission using forks, he could apply those results to both his multiple-harmonic and speaking telegraphs.[88]

Bell preferred a heuristic that combined substitution with holding constant, as Figure 18 illustrates.[89] This figure condenses a far more complicated sequence of experiments into two substitutions: Bell removed one electromagnet, holding all other aspects of the circuit constant, and then after several intervening steps introduced a dish of water into the circuit.[90] This dish of water served as a high-resistance medium between the two electrical contacts in the circuit. It is at this point that he opened a new slot, which we have labelled 'contacts', because his next line of experiments concerned the relationship between the surface area of the contacts immersed in the water. For example, at one point he immersed a bell in the water

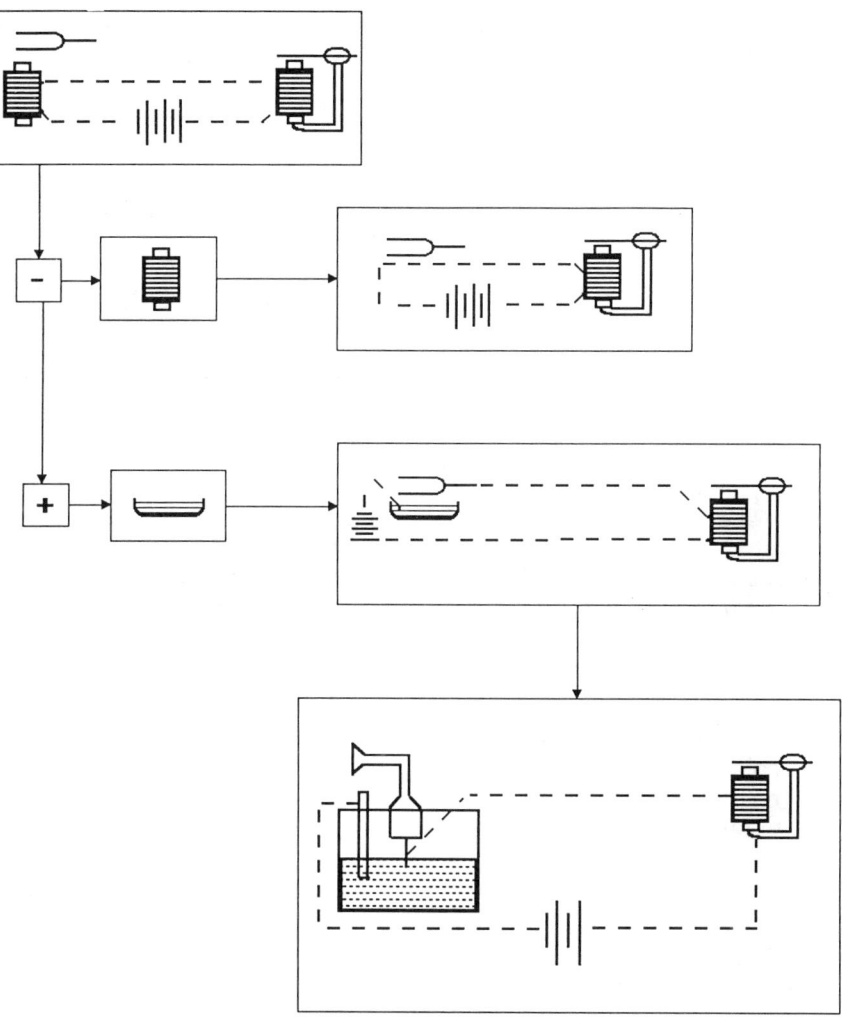

Figure 18 Map showing the two key substitutions that Bell made as he moved from experimenting with a tuning-fork and electromagnet to a speaking telegraph.

to see what happened when he maximized the area of the vibrating contact relative to the stationary one at the bottom of the dish. This result was not satisfactory, so he decided to try minimizing the size of the vibrating contact relative to the stationary one.[91]

This conclusion, after several variations, permitted Bell to transmit speech successfully on 10 March 1876. Bell spoke into a tube that ended in a diaphragm, causing the contact area between a needle mounted on the diaphragm and container of water to increase and decrease. An undulating current was produced and transmitted the words, 'Watson—Come here—I want you' to Bell's standard reed relay.[92] Watson came, and they knew they had succeeded.

Yet, as we will soon see, the introduction of water as a medium of high resistance once again brought Bell and Gray into conflict.

THE MECHANICAL TRANSMITTER, TUNED-BOX ANALYSER, LOVERS' TELEGRAPH AND GRAY'S CAVEAT

Figure 19 depicts another portion of Gray's multiple telegraph research, and we see that Gray conducted experiments using the same receivers as he employed in his experiments with the two-octave transmitter. Recall that each of these receivers was capable of receiving multiple tones. In the transmitter slot, however, he used a mechanical transmitter.[93] This device consisted of two heavy metal armatures resting on metal jumping cams on a belt-driven shaft. The pressure between the armatures and the cams was regulated by tension screws or by pressing a finger on the arma- tures. The operator thus could control the intimacy of contact between the armature and cams. Gray used this device to transmit a complex, groan- like sound to his receivers. After conducting these experiments, Gray realized that

> the mechanical transmitter confirmed what my previous experiments
> had lead me to believe: that not only could the receivers that had been
> named be used as receivers of articulate speech transmitted electrically,
> but that such speech could be transmitted through or from a single
> point. I mean by single point, without the intervention of a series of
> reeds or points differently tuned, and one that would be a common
> or universal transmitter, in the same sense that the receivers were
> universal or common.[94]

The idea of being able to transmit sounds and even speech through a single point surfaced again when Gray experimented with a recorder for his multiple telegraph[95] (see Figure 20). This device was designed to receive a message sent at a certain frequency from a multiple-harmonic transmitter such as the two-octave transmitter and then retransmit the signal to a multiple-telegraph recording device. At the end of the hollow box was another smaller box that had six wooden rods which fit inside the larger box. These rods allowed the smaller box to slide so the distance between the boxes could be adjusted. On one end of the smaller box, a parchment or gold-beater's skin diaphragm was attached. At the centre of this diaphragm was a platinum contact point. When a tone corresponding to the tuned resonant frequency of the reed and larger box was received, the whole apparatus vibrated and the smaller box and diaphragm vibrated in sympathy. The platinum contact point made and broke contact with a lever whose natural frequency was much less than that of the rest of the apparatus.[96] This whole arrangement was designed so that when the proper tone was received, it was heard audibly and recorded by a telegraph

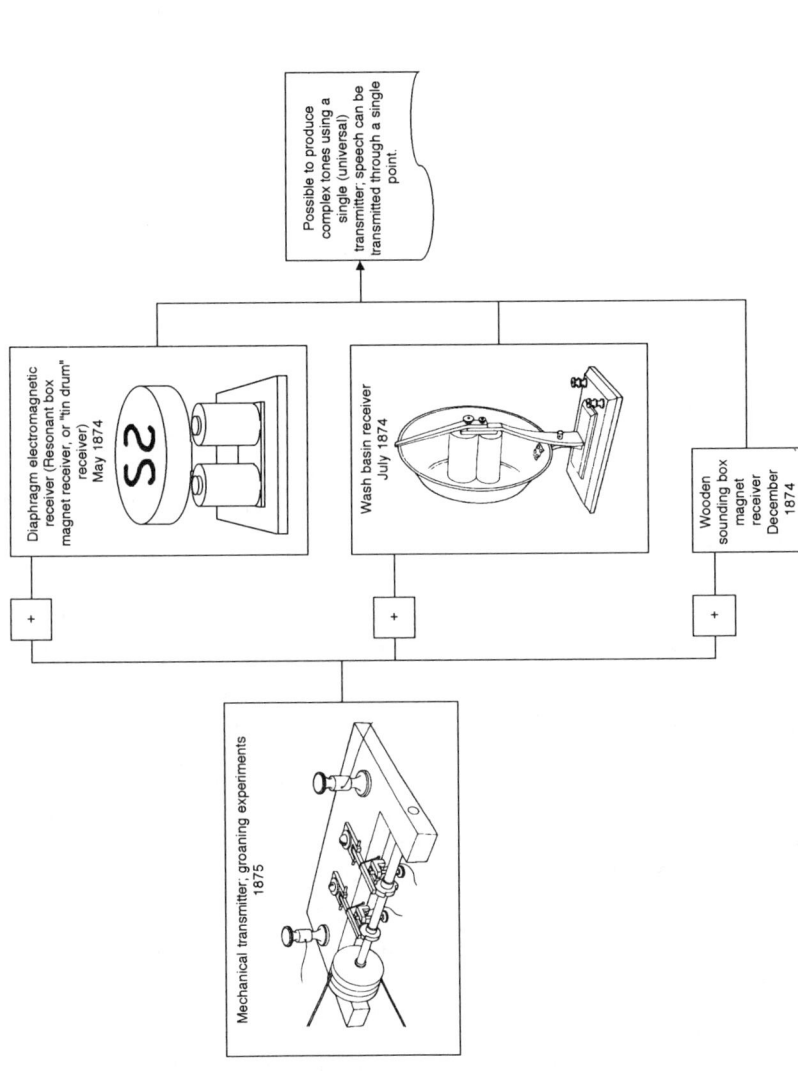

Figure 19 Gray's receiver combinations with his mechanical transmitter. Illustrations of mechanical transmitter and wash-basin receiver are from Hounshell, *op. cit.* (Figure 10), 159–60. Tin-drum receiver is from Gray patent, *op. cit.* (Figure 7).

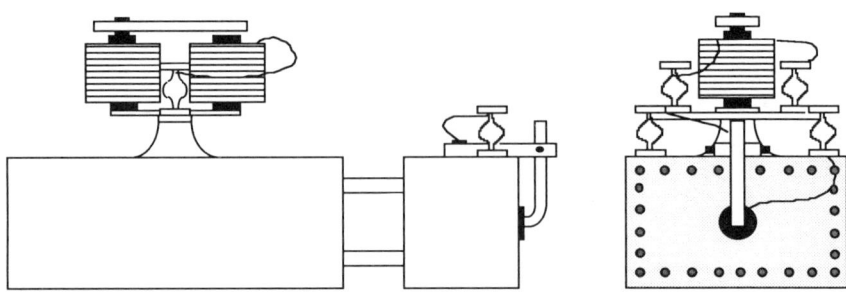

Figure 20 Gray's tuned-reed and box receiver with diaphragm and circuit-breaker. Redrawn from Elisha Gray, 'Local Circuit-Breakers for Electro-Harmonic Telegraphs', US Patent 194,671 (filed 15 February 1876, granted 28 August 1877).

recording instrument attached to the circuit controlled by the vibrating diaphragm.[97]

Although the platinum contact point and lever arm were designed to respond to a single tone, Gray observed the sensitive nature of the diaphragm, or 'drum-head' and contact-point lever-arm mechanism, or 'rattler':

> during these experiments I used several small drums, attaching the center of the drum-head to what I call a 'rattler,' designed to operate a local circuit. In experimenting with these drum-heads, I noticed that they were sensitive to noises made in the room from any source.
>
> These experiments were a stepping-stone which helped lead my mind in the direction it took when I afterwards saw the 'lovers' telegraph.'[98]

Figure 21 shows the lovers' telegraph, which became Gray's mental model for the transmission of speech. This device consisted of two resonant cavities, each with a diaphragm. A thread of one to two hundred feet was attached to both diaphragms and held taut. When one spoke into one resonant cavity, the sounds could be heard at the other end. Gray spoke of his reaction to this device in his deposition:

> [The lovers' telegraph] proved to my mind that all the conditions necessary for the transmission of an articulate word were contained in any single vibrating point . . . I saw that if I could reproduce electrically the same motions that were made mechanically at the center of the diaphragm . . . such electrical vibrations would be reproduced on a common receiver in the same manner that musical tones were.[99]

Thus, after he saw the lovers' telegraph, Gray had a clear mental model of how a speaking telegraph might function. His realization that speech could be sent by a transmitter through a single point, as he concluded from his mechanical transmitter experiments, was reinforced by the lovers' telegraph's use of the thread on a diaphragm. Combining this conclusion with his observation that his telegraphic receivers could be used for voice

Figure 21 The Lovers' Telegraph. *Source*: T. Du Moncel, *The Telephone, the Microphone, and the Phonograph* (New York, 1879), 34.

reception and mechanical representations from the lovers' telegraph and his receivers, Gray sketched his conception, the speaking telegraph, on 11 February 1876 (Figure 22). At that time, he was in Washington filing patents for several of his multiple-telegraph devices.

In this sketch, Gray designed a transmitter with a resonant speaking tube attached to a diaphragm. Beneath the diaphragm was a thin wire or rod, which was immersed in a high-resistance liquid, water. When one spoke into the resonant cavity, the diaphragm vibrated and changed the distance between the wire and a contact on the bottom of the water container. As the distance between these two contacts changed, the resistance, and hence the current, varied in the circuit. The changing current, which was sent over the line, energized the electromagnets in the receiver. The changing magnetic field attracted the metal diaphragm of the receiver by different amounts. The response of the receiver diaphragm was thus identical to the signal sent by the transmitter replicating the voice or other sounds spoken into the transmitter.

Gray assembled this speaking telegraph from mechanical representations he had used previously. The transmitter's resonant cavity and diaphragm were based on the resonant cavity of the lovers' telegraph. The dipping wire was inspired by the thread from the lovers' telegraph. The idea of using liquid variable resistance, Gray claimed, was 'old in the art at the time'.[100] Here Gray was perhaps referring to liquid-cell batteries in which the current of the batteries could be regulated by raising or lowering of the plates of the cells in and out of the electrolyte.

Gray's speaking-telegraph receiver resembled those that he had

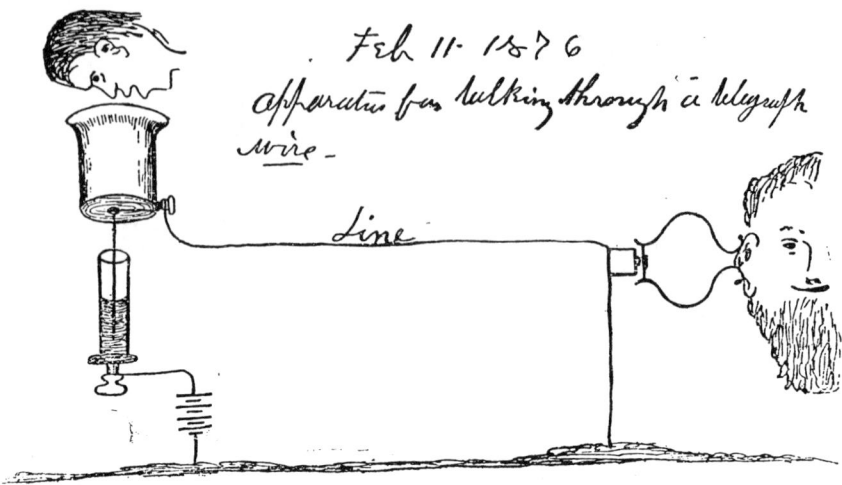

Figure 22 Gray's sketch for an 'apparatus for talking through a telegraph wire', 11 February 1876. *Source: The Telephone Appeals. Brief for American Bell Telephone Co.*, Supreme Court of the United States, October Term, 1886 (Boston, Alfred Mudge & Son, 1887).

employed in his multiple telegraph research. We still see the double-pole electromagnet, which came from the analysing receiver. The resonant cavity is similar to one which Gray used on a tuned receiver in December 1874.[101]

Gray's caveat

On the basis of his 11 February sketch, Gray submitted a preliminary patent application or caveat three days later in which he claimed 'the art of transmitting vocal sounds or conversations telegraphically through an electric circuit[102] (Figure 22). Because he did not have a working device, he filed a caveat or preliminary disclosure instead of a full application, and he was not especially concerned if some details of the apparatus were left somewhat vague.

The lovers' telegraph and the mechanical transmitter showed Gray that a complex sound such as the human voice could be transmitted through a single point. Yet in the caveat Gray did not emphasize this observation and instead he proposed employing multiple diaphragms just as he had used multiple transmitters in his harmonic telegraph: 'I contemplate, however, the use of a series of diaphragms in a common vocalizing chamber, each diaphragm carrying an independent rod, and responding to a vibration of different rapidity and intensity, in which case contact points mounted on other diaphragms may be employed.'[103] In his technical history of the telephone J.E. Kingsbury cited Gray's preference for multiple chambers to argue that in 1876 Gray was only at the level of understanding that Bell reached with his harp apparatus in 1874, in that each of these diaphragms would function like one of Bell's reeds and it would take a large number of them to reproduce the human voice.[104] But our framework shows that Gray realized only a single rod was necessary, and he may have thought multiple rods would improve the quality of the sound.

Bell's patent application

A few hours before Gray submitted his caveat, Hubbard had filed Bell's patent application for 'Improvements in Telegraphy'. Like Gray, Bell did not have a working speaking telegraph at this point. This fact led Gray's biographer, Lloyd Taylor, to claim that if Gray had submitted a patent instead of a caveat he would have been declared the inventor of the telephone.[105]

Given the vagaries of patent law maybe this could have happened. Bell's application, however, was far more detailed and thorough than Gray's caveat and reflected his general style: strong on theory but weak on working devices. Yet the application was breathtakingly broad: Bell claimed all communications applications of what he called an undulatory current. Therefore he discussed multiple telegraph and speaking telegraph applications. Unlike Gray, Bell saw these applications as distinct, though he illustrated both by using the same mechanical representation, his reed relay. For the telegraph he showed three stations, each with three reed relays functioning as transceivers. Although his description of the circuits was sketchy, Bell focused on the kind of current they would produce. For the speaking telegraph he attached two reed relays to a cone and diaphragm, one to act as transmitter, the other as receiver (see Figure 17). From a theoretical standpoint, this is elegant and simple. Bell devoted much of the application discussing the differences between undulatory and intermittent currents and why the former had advantages over the latter. Again, the discussion is quite theoretical, and Bell analysed sinusoidal curves in a Helmholtzian fashion.

Although Gray's speaking-telegraph transmitter would have produced an undulating current, Gray did not mention this in his caveat. Instead, Bell had a better grasp of the potential significance of this form of current and what could be gained by claiming it as an innovation, rather than trying to patent various possible combinations that could produce it. Bell sought to patent the theory behind a successful telephone and succeeded. In the long run, his bold claims made it hard for others to get around his patent.[106]

Upon receiving both Bell's and Gray's submissions, the Patent Office examined them and found them to be in interference. Bell's application was suspended for 90 days. The suspension was revoked on 25 February and Bell's solicitors were informed that the suspension was a misunderstanding of Bell's legal rights.[107]

Nevertheless, Bell visited the Patent Office on 26 February in order to discover the source of the interference. The patent examiner, Zenas F. Wilber, pointed to a passage in Bell's application that discussed a vibrating wire which dipped in and out of a high resistance liquid:

For instance, let mercury or some other liquid form part of a voltaic circuit, then the more deeply the conducting-wire is immersed in the mercury or other liquid, the less resistance does the liquid offer to the passage of the current. Hence, the vibration of the conducting-wire in mercury or other liquid included in the circuit occasions undulations in the current.[108]

Bell claimed he inserted these lines after the patent was submitted in order to clear up a potential interference with one of his own patent applications submitted previously.[109] This brief mention of variable resistance gave Bell priority over later liquid transmitters, like Gray's, which were submitted after his application.

Wilber's finger may have cued part of Bell's next line of research, for it was with a liquid transmitter that Bell first successfully transmitted speech (see Figure 18). Yet this does not mean Bell 'stole' the liquid transmitter from Gray; rather, Wilber's slip about the interference may have suggested a promising line of experiments to Bell. In the next section, we will compare Bells and Gray's liquid transmitters and show that they reflect the differences in their mental models.

Bell's liquid transmitter compared with Gray's caveat

The device Bell first used to transmit speech looks very similar to Gray's caveat (see Figure 23). This naturally has led both lawyers and historians to charge that Bell stole Gray's idea, perhaps through some chicanery at the Patent Office with Wilber.[110] Earlier, we showed how Bell and Gray arrived at similar vibratory circuit-breakers from different mental models. Furthermore, their mechanical representations show subtle differences characteristic of their mental models. In this section, we will show how Bell and Gray could have arrived at similar combinations of mechanical representations for a liquid transmitter from different mental models and, furthermore, how the superficial similarities of these devices mask important differences.

Let us begin with a close look at these devices in Figure 23. First and most obviously, each inventor used a different mechanical representation as a receiver. Gray used an electromagnet-resonant chamber combination reminiscent of his wash-basin and other receivers. Bell used one of his standard reed relays. Second, there are subtle but critical differences in the transmitters. In Gray's device, it is the *distance* between the wire and the contact at the bottom of the liquid that changes when one speaks. Gray's wire was therefore immersed very deeply in the water, and the contact with the rest of the circuit lies directly beneath it. In Bell's instrument, it is the *surface area* of the needle suspended in water that changed.[111] The needle from his diaphragm lay close to the surface of the liquid, and the other contact was off to one side, in a manner similar to the spark arrester he attempted to patent in January 1876.[112] Indeed, one could argue that Bell substituted his spark arrester mechanical representation into the 'contacts slot'.

Further evidence that Bell investigated surface area, not depth, comes

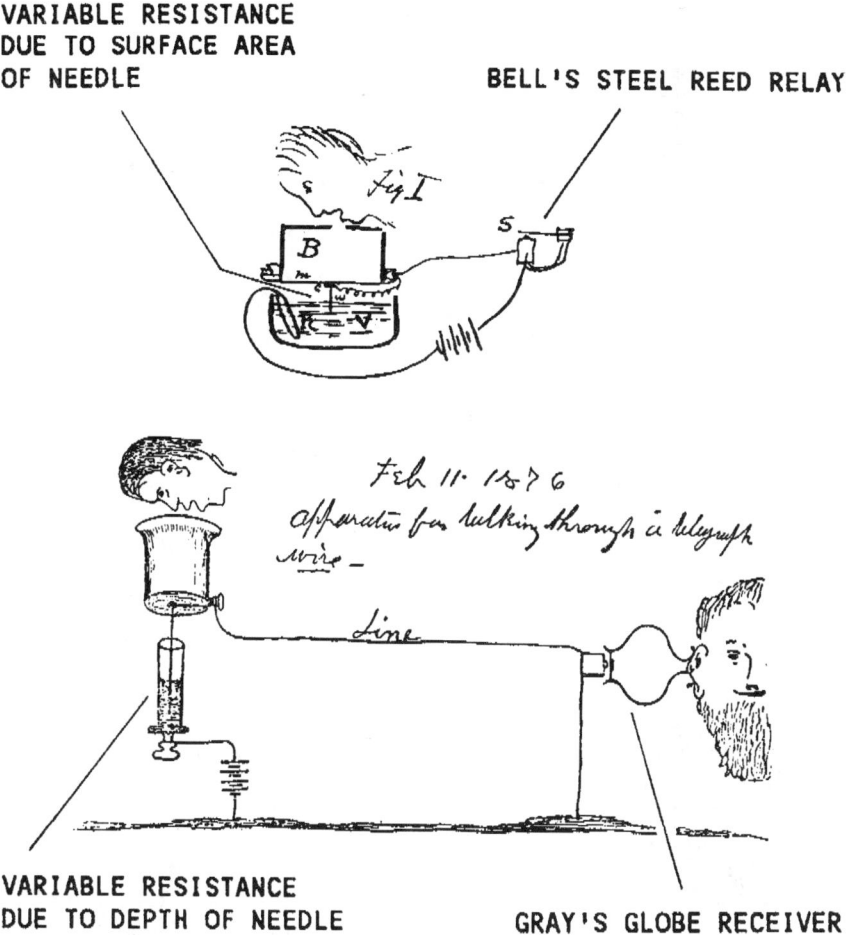

VARIABLE RESISTANCE
DUE TO SURFACE AREA
OF NEEDLE

BELL'S STEEL REED RELAY

VARIABLE RESISTANCE
DUE TO DEPTH OF NEEDLE

GRAY'S GLOBE RECEIVER

Figure 23 Bell and Gray's liquid transmitters compared. The top sketch is Bell's first drawing of his idea for a liquid variable-resistance transmitter. Note that the two contacts were side by side and that the area of the vibrating contact w was much smaller than the area of the contact R. Bell also used one of his reed relays as a receiver. The bottom sketch shows Gray's liquid variable-resistance transmitter. Note the long needle dipping in the water, which almost touched the contact at the bottom. Gray used one of his standard mechanical representations as a receiver. Bell sketch is from Bell notebook, *op. cit.* (Figure 16), 39.

from an experiment that led to his successful transmission of speech (Figure 24). In one test, he replaced the vibrating contact in his liquid transmitter in order to see what happened when he made the surface area of the vibrating contact as large as possible. When the bell in this circuit was rung, no sound was heard from the reed receiver. From this and other experiments Bell concluded that the sound from the receiver was loudest

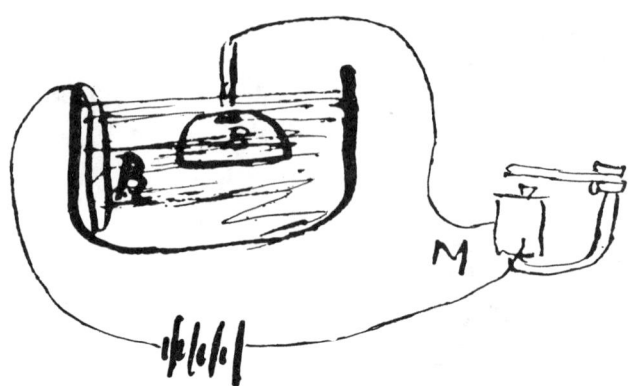

Figure 24 Bell's attempt to maximize the area of the vibrating contact by substituting a brass bell B for the tuning-fork or vibrating contact. R was a ribbon of brass and M was a reed relay. From Bell notebook, *op. cit.* (Figure 16), 38.

when the surface area of the side contact was largest and the vibrating surface in contact with the water was smallest. This conclusion is reflected in his successful liquid transmitter, which has a vibrating needle barely making contact with the water and a large brass pipe as the other contact.[113]

Another difference in the transmitters is the devices into which one speaks. Gray shows an insulated cylinder of glass; Bell shows a speaking tube of the sort he used in his experiments with a manometric flame capsule. Finally, Gray contemplated using a series of liquid transmitters. Gray may have thought several diaphragms would be necessary because his mental model of a multiple telegraph presumed multiple transmitters. It may be that he thought transmission would be improved by a series of diaphragms sensitive to different pitches. To summarize, Bell's liquid transmitter and Gray's caveat device stem from different mental models and include unique mechanical representations. It is certainly possible that Wilber, the patent examiner, encouraged Bell to pursue liquid variable resistance when he pointed to the lines in Bell's patent that conflicted with Gray. But Bell did not simply copy Gray's transmitter, as some have suggested.[114]

Indeed, the differences go even deeper. Part of the controversy had to do with the fact that each inventor had a fundamentally different mental representation of what constituted a telephone. As Bell wrote to Gray on 2 March 1877,

> I have not generally alluded to your name in connection with the invention of the Electric 'Telephone,' for we seem to attach different significations to the word. I apply the term only to an apparatus for transmitting the voice (which meaning is strictly in accordance with the derivation of the word) whereas you seem to use the term as expressive of any apparatus for the transmission of musical tones by the electric current.[115]

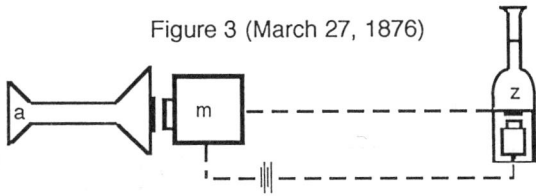

Figure 3 (March 27, 1876)

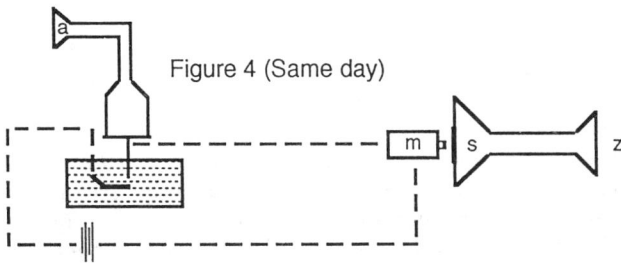

Figure 4 (Same day)

Figure 25 Bell's experiments with induction (Figure 3) and variable resistance (Figure 4) telephones, 27 March 1876. One could speak into either a or z in both devices; s was a clock spring and m an electromagnet. In Figure 4, the needle vibrated in water. Redrawn from Bell notebook, *op. cit.* (Figure 16), 85–6.

Gray responded by giving Bell 'full credit for the talking feature of the telephone'. This emphasis on talking as a feature suggested the extent to which Gray regarded talking as only one relatively minor function of the telephone invention.[116]

BELL'S WORK AFTER THE PATENT

A full account of Bell's telephone researches after his successful liquid transmitter lies beyond the scope of this paper. What is important to note here is that he returned to a design very similar to his patent in the summer of 1876. At this point, Bell replaced his liquid transmitter with an instrument which used electromagnetic induction, much like his reed relay. After testing Bell's liquid transmitters, Bernard S. Finn concluded that Bell abandoned his liquid instruments because he was not able to get them to work as well as his magneto devices.[117] But Bell's experimental notebooks include numerous liquid transmitter designs that worked well, according to his own comments, and many of his magneto devices did not perform up to his expectations. This is illustrated by two experiments Bell tried seventeen days after his first successful transmission of speech (Figure 25). Of the experiment in Figure 3, Bell wrote, 'Upon singing into A sounds were heard at Z and upon singing into Z sounds were heard at A much more distinctly than in any of the preceding cases.' Of Figure 4, he wrote,

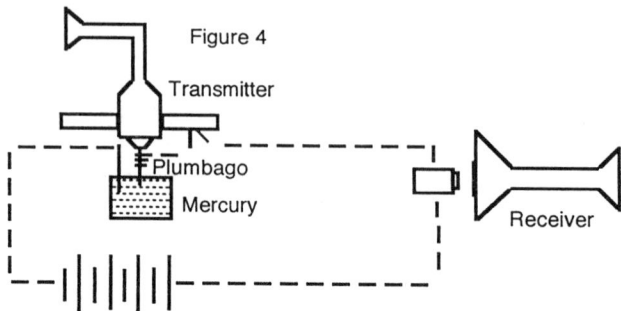

Figure 26 Bell's variable-resistance transmitter with a plumbago contact in a mercury cup, 2 April 1876. From Bell notebook, *op. cit.* (Figure 16), 94.

'On speaking into A sounds were perfectly audible from Z. Much louder than any yet obtained with the voice. Unmistakably articulate sounds proceeded from Z.'[118] There is no indication here that the liquid transmitter experiments were producing poorer results than the magneto ones. In both cases, however, Bell was still having trouble with consonants, although vowels came through clearly.

A few days later, Bell revealed the theoretical importance of these experiments. Figure 26 shows a transmitter in which a plumbago point vibrates in mercury. The receiver was one of Bell's typical magneto devices. When the plumbago was only slightly immersed in the mercury, the sound at the receiver was feeble, but then suddenly it would burst forth. At such times Bell noticed that a bright spark was visible between the plumbago and the mercury, showing that the plumbago had vibrated in and out of mercury, creating an intermittent or make-or-break current. Bell noticed a significant difference in the quality of the sounds produced by the intermittent and undulatory currents. As long as the plumbago never left the mercury, Bell heard not only the pitch but also the quality, or the timbre, of the speaker's voice. When the spark appeared at plumbago, only the pitch of the voice could be discerned, not the quality. This led Bell to conclude 'that my theory is correct—that musical notes which conflict with one another when transmitted by means of an intermittent (current) *will not interfere with one another when the undulatory current is employed*'.[119] For Bell, this conclusion meant that the undulatory current could be used both in a telegraph and in a telephone, and it reinforced Bell's preference for the undulating current.

On 18 November 1876, Bell wrote his final conclusion about the problems with liquid transmission:

> as the wire proceeds into the liquid the current becomes absolutely stronger and stronger, but as the membrane which carries the wire descends lower and lower its motion becomes slower and slower, until it stops. Now if the current were purely undulatory its intensity would diminish as the motion of this wire diminished and when it reached

its lowest point it would be zero, but the fact is that at the lowest point the current reaches its minimum and hence the vibration must be very greatly distorted, probably still more distorted than in the case of the induced undulatory current.[120]

In other words, Bell had theoretical as well as practical reasons for abandoning the use of variable resistance in his telephones, even though both Gray and Edison subsequently used this phenomenon to create their own practical telephone.[121] Instead, Bell's first production telephones follow his original ear mental model very closely. For the ossicles, he substituted a horseshoe magnet. When one spoke into a metal diaphragm, an undulating electric current was generated in an induction coil wrapped around the ends of the magnets.[122]

GRAY'S WORK AFTER HIS CAVEAT

Gray did not pursue the development of the telephone immediately after he submitted his patent caveat. Several of his backers reinforced his attitude that the telephone was simply a toy, with no grand commercial payoff.[123] Consequently, after submitting his caveat Gray did nothing on the telephone for six months. In June or July 1876, Gray constructed a model of the speaking telegraph. The transmitter functioned identically to the one described in his caveat. Following his combination heuristic, he tried different receivers in combination with this transmitter. These experiments set up a pattern that dominated his future telephone efforts: he attempted to patent a combination that would block Bell.[124]

In October 1877, Gray filed two patent applications to challenge and limit the scope of Bell's telephone patent.[125] Figure 27 shows that in both these applications, Gray took on Bell's ideas of using electromagnetic induction and having identical transmitters and receivers. However, Gray employed his own mechanical representations in both of these attempts. In one application he carried over the receiver he had designed in his February 1876 caveat, while in the other he used his wash-basin receiver (see the top two drawings in Figure 27). With these applications, Gray also sought to obtain a broad patent covering multiple-harmonic telegraphy in order to convince the Patent Office that Bell's telephone was a subclass already governed by principles covered by harmonic multiple telegraphy. To prove this point, Gray described a system using a multiple-tone transmitter and his wash-basin receiver (as shown in the bottom drawing in Figure 27). In submitting an application for the principles underlying his multiple-harmonic telegraph, Gray may well have been imitating Bell's strategy of broadly patenting the use of undulatory currents in multiple and speaking telegraphs.

Both of Gray's October 1877 applications were rejected by the Patent Office. Gray eventually obtained several telephone patents, most of which were, ironically, improvements of interchangeable induction transmitter/receiver devices, such as the bipolar telephone.[126]

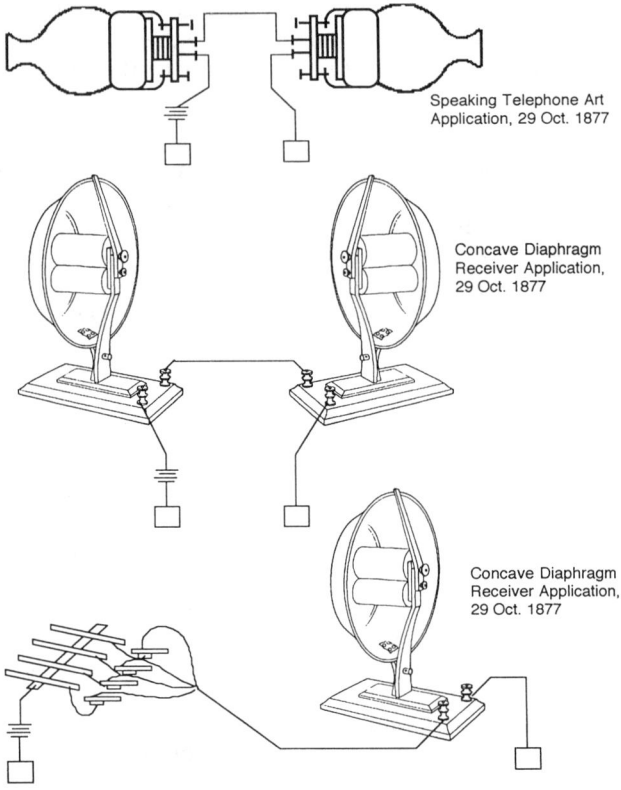

Speaking Telephone Art
Application, 29 Oct. 1877

Concave Diaphragm
Receiver Application,
29 Oct. 1877

Concave Diaphragm
Receiver Application,
29 Oct. 1877

Figure 27 Gray's attempts to patent around Bell. Redrawn from Elisha Gray, 'Art of Transmitting Vocal Sounds Telegraphically', US Patent Art application filed 29 October 1877 and 'Improvements in Electric Telephony', US Patent application filed 29 October 1877, in *Bell Telephone Company et al. v. Peter A. Dowd*, Telephone Suits, US Circuit Court, Massachusetts District (Boston, Alfred Mudge & Son, 1880), Part II. Illustration of wash-basin receiver is from Hounshell, *op. cit.* (Figure 10), 159.

CONCLUSION

Our framework has revealed important differences in the mental models, mechanical representations and heuristics of two inventors whose devices often appeared so similar that each was convinced the other had stolen part of his idea for either the multiple-harmonic telegraph or the telephone. In fact, each inventor followed a unique path to the speaking telegraph, even though their paths intersected at several points and each influenced the other.

Klahr and Dunbar identify two different styles in scientific thinking: theoretical and experimental.[127] Bell made a deliberate decision to pursue theory because of his lack of electrical knowledge and his strong background in speech.[128] Consequently, he concentrated on developing powerful mental models, and he settled for a limited stock of mechanical representations sufficient to confirm his understanding of how a speaking telegraph could be built. His heuristics included a combination of substitution and holding constant to generate new mechanical representations.

Gray, in contrast, generated a far richer store of mechanical representations and used a 'try all combinations' heuristic to arrive at new inventions. His mental models were based closely on existing devices like the Reis telephone and the lovers' telegraph. Thus, Gray comes closer to the experimental style proposed by Klahr and Dunbar, though he was not an experimental scientist; he was a practical inventor whose innovations were linked closely to his 'hands-on' experience.[129]

A whig conclusion would be to argue that Bell's theoretical style proved superior to Gray's practical one, just as the theoretical scientist of the twentieth century has supposedly triumphed over the empirical inventor of the nineteenth century. Yet such a conclusion is not warranted. Bell did not succeed in patenting a speaking telegraph just because he had a good theoretical conception of what he was doing. His ambitious father-in-law, the vagaries of patent law and other factors played very important roles. Similarly, Gray's backers did not encourage him to pursue the speaking telegraph. Nevertheless, Gray was a very creative inventor whose methods also deserve investigation. A cognitive framework does not allow us to determine which of two inventors was 'better'. Instead, it helps us appreciate what was unique about each inventor's style. Likewise, a cognitive framework cannot settle questions about simultaneous invention; just as the mantle of inventor is granted as a result of complex negotiations among different interests and communities, so the label 'simultaneous inventor' is also granted. What a cognitive framework can do is reveal important similarities and differences in the processes by which artefacts and ideas are created.

Acknowledgements
This research was supported by the History and Philosophy of Science and Technology Program of the National Science Foundation (Grants DIR-8722002 and 9012311), the Spencer Foundation, and the School of Engineering and Applied Science. We wish to thank Nancy Briscoe, Robert Bruce, Kim Candee, Bernard S. Finn, Sheldon Hochheiser, Keith Nier, Christi Nilsen, Ron Overmann, and Jeffrey Sturchio for their comments and assistance. A preliminary version of this paper was presented at the annual meeting of the American Association for the Advancement of Science on 16 February 1991 in Washington, DC.

Notes and References
1. Robert V. Bruce, *Bell: Alexander Graham Bell and the Conquest of Solitude* (Little, Brown, 1973), 164–76; David A. Hounshell, 'Elisha Gray and the Telephone: On the Disadvantages of Being an Expert', *Technology and Culture*, 1975, 16: 133–61.

2. Bernard S. Finn, 'Alexander Graham Bell's Experiments with the Variable-Resistance Transmitter', *Smithsonian Journal of History*, 1966, 1: 1–16.

3. On 14 January 1901, Gray wrote a letter to the *Electric World and Engineer* in which he argued that he had eventually 'became convinced . . . that I had shown

[Bell] *how* to construct the telephone with which he obtained his first results'. See Lloyd W. Taylor, 'The Untold Story of the Telephone', unpublished manuscript, Box 6, Folder 6, Elisha Gray Papers, National Museum of American History, Smithsonian Institution, Washington, DC. (Hereafter this collection is cited as GP.) Indeed, throughout their multiple telegraph competition such charges periodically surfaced. For example, on 26 April 1875, Elisha Gray wrote to his attorney A.L. Hayes, 'I have read Bell's claim and it seems to me he could not have described my invention better if he had copied it.' The claim here is not from Bell's telephone patent, but from one of his multiple telegraph patents, presumably 161,739, filed 6 March 1875. As early as October 1874, Bell was making similar statements about Gray: 'I believe that Mr. Adams [Bell's solicitor] had unintentionally, by the mention of my invention, given Mr. Hayes a hint which at once sent Mr. Gray upon my track'. See Alexander Graham Bell, *The Multiple Telegraph* (Boston, Franklin Press, Rand, Avery & Co., 1876), 15 in Telegraph-Miscellaneous Fol., Box 274, Bell Family Papers, Manuscript Division, Library of Congress, Washington, DC. (Hereafter this collection will be cited as BFP.)

4. For a summary of this litigation, see Frederick Leland Rhodes, *Beginnings of Telephony* (Harper & Brothers, 1929), 49–75 and 207–24.

5. Karl Popper was especially vehement in his view that there was a method to how scientists justified their hypotheses, but that there was 'no such thing as a logical method of having new ideas, or a logical reconstruction of this process.' See his *Logic of Scientific Discovery* (Basic Books, 1959), 31–2. This separation of discovery and justification has come to shape much of the scholarship in the social history and sociology of science and technology. For a discussion of how discovery and justification were separated see Augustine Brannigan, *The Social Basis of Scientific Discoveries* (Cambridge University Press, 1981), 3–6. For a particularly strong statement of the importance of focusing on how scientists justify their statements see Bruno Latour, *Science in Action: How to Follow Scientists and Engineers through Society* (Open University Press, 1987).

6. On the contextual approach to the history of technology consult J.M. Staudenmaier, *Technology's Storytellers: Reweaving the Human Fabric* (MIT Press, 1985), 162–201 and Merritt Roe Smith and Steven C. Reber, 'Contextual Contrasts: Recent Trends in the History of Technology', in S.H. Cutcliffe and R.C. Post, eds, *In Context: History and the History of Technology* (Lehigh University Press, 1989), 133–49. On the social constructivist approach consult Trevor J. Pinch and Wiebe E. Bijker, 'The Social Construction of Facts and Artifacts: Or How the Sociology of Science and the Sociology of Technology Might Benefit Each Other', in W.E. Bijker, T.P. Hughes and T.J. Pinch, eds, *The Social Construction of Technological Systems: New Directions in the Sociology and History of Technology* (MIT Press, 1987), 17–50.

7. Some social constructivists are sensitive to this issue. Steve Woolgar, for example, criticizes sociological as well as cognitive reductionism; see S. Woolgar, 'Reconstructing Man and Machine: A Note on Sociological Critiques of Cognitivism,' in Bijker, Hughes and Pinch, *op. cit.* (6), 311–28.

8. See Hounshell, *op. cit.* (1), 'Bell and Gray: Contrasts in Style, Politics, and Etiquette', *IEEE Proceedings*, 1976, 64: 1305–14, and 'Two Paths to the Telephone', *Scientific American*, January 1981, 244: 156–63.

9. Mental representation is the central problem in cognitive science. How can we rigorously investigate what cannot be directly observed? This was the problem that originally led behaviourists to focus exclusively on observable human actions, and may lie behind the insistence of some sociologists that all we can deal with is 'inscriptions'. For a lively debate on this and related issues, see the symposium on

'Computer Discovery and the Sociology of Scientific Knowledge', *Social Studies of Science*, 1989, 19: 563–696. See also M.E. Gorman, *Simulating Science: Heuristics, Mental Models and Technoscientific Thinking* (Indiana University Press, 1992).

10. For a discussion of the importance of linking cognitive and social approaches to science and technology studies, see Michael E. Gorman and W. Bernard Carlson, 'Can Experiments Be Used to Study Science?' *Social Epistemology*, 1989, 3: 89–106. Lest one think that we are opposed to a social analysis of technology, we would point out that one of us has contributed to the social history of technology; see W. Bernard Carlson, *Innovation as a Social Process: Elihu Thomson and the Rise of General Electric, 1870–1900* (Cambridge University Press, 1991).

11. Ryan D. Tweney makes a distinction between framework and theory:

Truth claims in a theory are based on the familiar strategies of scientific practice, while truth claims in a framework rely on interpretive procedures more akin to the methods of historical scholarship. A theory is an attempt to construct a model of the world which meets certain criteria of testability; it makes predictions, is potentially disconfirmable, and has interesting consequences. A framework is an attempt to *re*-construct a model of the world which meets criteria other than testability as such. An adequate framework is one that is consistent with the details of the process, is interestingly related to our theories of the world, and reduces the apparent complexity of the real world process in a way which permits anchoring the framework to the data. In effect, an adequate framework must allow us to see order amid chaos.

See his 'Framework for the Cognitive Psychology of Science', in B. Gholson, W.R. Shadish, R.A. Niemeyer and A.C. Houts, eds, *Psychology of Science* (Cambridge University Press, 1989, 344). Our framework and maps allow us to perceive order among the chaos of the inventor's sketches, artefacts, caveats and notes. Our framework might eventually evolve into a more formal, theoretical model, but for now it provides a rigorous basis for comparing inventors.

12. Quoted in W. Bernard Carlson and Michael E. Gorman, 'Understanding Invention as a Cognitive Process: The Case of Thomas Edison and Early Motion Pictures, 1888–1891', *Social Studies of Science*, 1990, 20: 387–430, 396. This article includes a detailed discussion of how this cognitive framework illuminates the invention of the kinetoscope.

13. For good reviews, see D. Gentner and A.L. Stevens, *Mental Models* (Lawrence Erlbaum Associates, 1983), 99–129 and W.B. Rouse and N.M. Morris, 'On Looking Into the Black Box: Prospects and Limits in the Search for Mental Models', *Psychological Bulletin*, 1986, 100: 349–63. Ronald Finke deals with the role of visualization in invention by asking experimental subjects to mentally construct new devices out of simple shapes. In effect, these subjects are creating and manipulating mental models, though Finke does not cite this literature, nor does he include comparative case-studies detailing the processes of actual inventors. See R. Finke, *Creative Imagery: Discoveries and Inventions in Visualization* (Lawrence Erlbaum Associates, 1990).

14. This term is adapted from R.J. Weber and D.N. Perkins, 'How to Invent Artifacts and Ideas', *New Ideas in Psychology*, 1989, 7: 49–72.

15. We have developed the concept of mechanical representations primarily through a close study of the work of several inventors, but we have also been inspired by several historians of technology. Reese V. Jenkins suggested that Edison frequently employed certain mechanical and electrical components (such as the cylinder and stylus) in his inventions. See his article, 'Elements of Style: Continuities in Edison's Thinking', *Annals of the New York Academy of Sciences*, 1984, 424: 149–62. To some extent, mechanical representations resemble the technical structures which Bertrand Gille defined as the basic tools and elements underlying all technology. See his *History of Techniques* (Gordon & Breach, 1986), vol. 1, 10–14.

Finally, Eugene Ferguson has described how several technologists catalogued mechanical movements; one example is the mechanical alphabet created in the eighteenth century by the Swedish engineer Christopher Polhem. See his article, 'The Mind's Eye: Nonverbal Thought in Technology,' *Science*, 1977, 197: 827–36.

16. For a discussion of how Edison used this mechanical representation in the kinetoscope, consult Carlson and Gorman, *op. cit.* (12), 398–9. For examples of how he used it throughout his telegraph work, see Reese V. Jenkins *et al.* eds, *The Papers of Thomas A. Edison*, vol. 1: *The Making of an Inventor, February 1847–June 1873* (Johns Hopkins University Press, 1989), especially 200, 323–5, 370, 407 and 428.

17. Cognitive psychologists have studied scientific heuristics in detail, using experimental, computational and historical methods; see Gorman, *op. cit.* (9). Very little attention has been paid to heuristics used by inventors; see Weber and Perkins, *op. cit.* (14) for an exception.

18. Carlson and Gorman, *op. cit.*, (12), 400–1.

19. Gordon Hendricks, *The Edison Motion Picture Myth* (University of California Press, 1961); Robert Skylar, *Movie-Made America: a Social History of American Movies* (New York, Random House, 1975), 10–11.

20. For further discussion of our mapping methodology, see Michael E. Gorman, Christy Nilsen and W. Bernard Carlson, 'A New Method for Mapping the Invention Process', in preparation for *Experience and Technique: How Science Practice Transforms Science Experience*, ed. David Gooding (University of Chicago Press, forthcoming).

21. Robert L. Thompson, *Wiring a Continent: The History of the Telegraph Industry in the United States, 1832–1866* (Princeton University Press, 1947), 421–6; Jenkins *et al.*, *op. cit.* (16), Vol. 1, 13 and 101.

22. Bruce, *op. cit.* (1), 93 and 126–7.

23. Bell, *op. cit.* (3), 3.

24. Hounshell, *op. cit.* (8), makes the case that Bell deliberately sought to establish strong links to the scientific community of the day. Bell's careful attention to science and scientists was one of the reasons for his success.

25. Bruce, *op. cit.* (1), 42–3 and 47–8. Ellis translated Helmholtz's treatise on acoustics into English; see Hermann L. Helmholtz, *On the Sensations of Tone as Physiological Basis for the Theory of Music*, trans. Alexander J. Ellis (Longmans, Green, 1875).

26. Alexander Graham Bell, *The Bell Telephone* (Boston, American Bell Telephone Co., 1908), 9.

27. Bell was unsure about the exact date and thought this experiment occurred in January or October of that year; see Bell, *op. cit.* (3), 6.

28. One should note that the fork that dipped in mercury was in a local circuit. This provided a constant intermittent current in the transmitter so that its fork was in constant vibration.

29. Bell, *op. cit.* (3), 6.

30. J. Baille, *The Wonders of Electricity*, trans. J.W. Armstrong (Scribner, Armstrong, 1872), 142–3.

31. Bell, *op. cit.* (3), 14. Bell eventually patented the general principle of using rotating magnets to induce a continuous current in a closed circuit; see A.G. Bell, 'Generating Electric Currents,' US Patent No. 181,553 (filed 12 August 1876, granted 29 August 1876).

32. These experiments are described in a letter from A.G. Bell to Gardiner Hubbard on 27 November 1874. BFP, Box 80.

33. Between 1860 and 1864, Reis built and demonstrated a series of electrical devices that transmitted and received musical notes, which he called his 'telephone'.

Although Reis produced several designs, his transmitter generally consisted of a membrane diaphragm with a metal contact on its surface. If one sang or spoke, into the diaphragm, the metal contact moved up and down and touched a needle suspended over the diaphragm. Because the diaphragm contact and the needle were in an electrical circuit, they functioned as a switch and created rapid intermittent pulses which were sent to a receiver. The receiver was an iron needle wrapped in a helix of wire. The electrical pulses rapidly magnetized and demagnetized this receiver, reproducing the pitch of a tone being transmitted. For more details, see Michael E. Gorman and W. Bernard Carlson, 'Interpreting Invention as a Cognitive Process', *Science, Technology, and Human Values*, 1986, 15: 131–64, 135 and Silvanus P. Thompson, *Philipp Reis: Inventor of the Telephone* (E. & F. Spon, 1883).

34. Bell, *op. cit.* (26), 43.

35. Bell, *op. cit.* (3), 16.

36. A detailed consideration of these slots and sub-slots lies beyond the scope of this paper; therefore we will describe Bell's experiments designed to improve the circuit connections in a general way.

37. Bell, *op. cit.* (3), 8.

38. Bell, *op. cit.* (26), 16.

39. Bell, *op. cit.* (26), 18–19.

40. *Ibid.* 20–1.

41. *Ibid.* 34.

42. Bell, *op. cit.* (3), 14–15.

43. Elisha Gray, 'Telegraph-Relay Instrument', US Patent 69,424 (granted 1 October 1867); 'Improvement in Telegraph-Repeaters', US Patent 114,938 (granted 16 May 1871); and 'Improvement in Printing-Telegraph Instruments', US Patent 132,907 (granted 12 November 1872).

44. Gray had employed the vibrating electrotome in previous experiments. In the winter of 1866–7, Gray connected a vibrating electrotome in a circuit with a polar relay receiver. In this experiment Gray configured the vibrating electrotome so that it produced an undulatory current which had a frequency in the audible range. The polar relay responded at this same audible frequency, and Gray realized that musical tones could be sent over a telegraph line. In effect this experiment gave him an initial mental model for transmitting tones across a telegraph wire. The components of the polar relay and electrotome also become key mechanical representations in the construction of Gray's later transmitters and receivers. See Elisha Gray, *Experimental Researches in Electro-Harmonic Telegraphy and Telephony: 1867–1878* (New York, Russell Brothers, 1878; reprinted in George Shiers, ed., *The Telephone: An Historical Anthology*, Arno, 1977), 10.

45. The vibrating electrotome also played a role in Gray's decision to use animal tissue in the circuit. The vibrating electrotome was typically used as a device for administering low-level shocks, for either medical purposes or novelty. Thus, Gray's mental model of the animal tissue experiments came from his experience with using the vibrating electrotome as a low-level shock device. See [Daniel Davis], *A Manual of Magnetism* (Boston: by the Author, 1852), 280.

46. In his original patent application, executed 20 June 1874, Gray discussed the circuit-breaking properties of the Reis transmitter. Gray designed each circuit-breaking electromagnet in his two-tone transmitter to be a more reliable method of generating musical tones than Reis's method of having someone sing a pitch near the diaphragm fitted with a mercury cup circuit-breaker. Thus Gray's mental model of his two-tone transmitter was based on the principles of the Reis telephone transmitter. See Gray's 'Magnet Receiver Application', in *Bell Telephone Company et al. v. Peter A. Dowd*, Telephone Suits, US Circuit Court, Massachusetts District

(Boston, Alfred Mudge & Son, 1880), Part II, 583–7. Gray, however, actually constructed his two-tone transmitter from mechanical representations obtained from the vibrating electrotome. These mechanical representations include the oscillating armature, the use of electromagnets with primary and secondary windings, and a rigid platinum contact point at a point on the armature. See Davis, *op. cit.* (45), 271–81.

47. Elisha Gray, 'Electric Telegraph for Transmitting Musical Tones', US Patent 166,096 (filed 19 January 1875, granted 27 July 1875).

48. *Bell Telephone v. Dowd, op. cit.* (46), Part I, 113.

49. *Bell Telephone v. Dowd, op. cit.* (46), Part I, 78. See also 'List of Exp. App' 214,294, GP.

50. Elisha Gray, 'Improvement in Transmitters for Electro-Harmonic Telegraphs', US Patent 165,728 (filed 28 June 1875, granted 20 July 1875). See also *Bell Telephone v. Dowd, op. cit.* (46), Part I, 113.

51. Gray, *op. cit.* (44), 20–1.

52. Gray, 'Magnet Receiver Application', in *Bell Telephone v. Dowd, op. cit.* (46), Part II, 583–7.

53. Gray used a violin as a receiver in some of his early experiments. The only difference between the violin-receiver resonant cavity and the tin-box resonant cavity was a variation of material. The violin receiver was based, however, on the principle of rubbing animal tissue on a conducting surface. The connection between the tin-drum receiver and the violin receiver was that the resonant chambers functioned as interchangeable mechanical representations, not that the two receivers are based on the same mental model. See *Bell Telephone v. Dowd, op. cit.* (46), Part I, 72.

54. Gray was familiar with the Reis telephone and knew that it was nearly capable of transmitting the human voice across, although its success was limited to tones generated by the human voice, not actual speech. See Gray, *op. cit.* (52). The voice-like sounds generated by the spark were perhaps more convincing to Gray than written accounts of the Reis device.

55. Bell, *op. cit.* (3), 15.

56. Quotes are from 'Defendant's Exhibit Bell Caveat, W.P. Preble, Jr., Examiner', in *Bell Telephone v. Dowd, op. cit.* (46), Part I, 708 and 713 respectively.

57. Bell, *op. cit.* (3), 15.

58. In a letter to his parents on 5 February 1875 (Box 5, BFP), Bell summarized his concerns: 'I have been suspicious of Mr. Adams from the moment that he offered to find out for me what Mr. Gray was doing—and I have been careful to avoid letting Mr. Adams know of my latest experiments—as I thought if he could so easily get information for me from Mr. Gray's solicitor Mr. G. could do the same through him.' This letter suggests the extent to which both Bell and Gray's attorneys passed information to their clients. These attorneys were not above trying to 'play both sides of the street', either; in the same letter, Bell mentions that Adams is threatening to work for Gray, and 'He knows all my *dates*, and has some of my papers in his possession now.'

59. Gardiner Hubbard to Alexander Graham Bell, 15 August 1874, Box 80, BFP.

60. Gardiner Hubbard to Alexander Graham Bell, 18 November 1874, Box 80, BFP.

61. We have been unable to locate either of these applications, though their substance is described in other places. For Bell's, see *op. cit.* (26), 40 and 42. For Gray, see *op. cit.* (44), 41.

62. We have been unable to locate Gray's patent application dated 23 February 1875. A description of the device in that application, however, is included

in E. Gray, 'Improvement in Receivers for Electro-Harmonic Telegraphs', US Patent 166,094 (filed 28 June 1875, granted 27 July 1875). The original device's contact lever was not elbow-shaped, nor did it utilize a regulating spring for the contact lever.

63. These included his tin-plate, violin, tin-drum, tambourine and wash-basin receivers.

64. Gray, *op. cit.* (44), 33–4. The story of how Gray invented his analysing receiver has been glossed over several times in technical histories of the telephone and telegraphy. For instance, in one of his histories of the telegraph and telephone, George B. Prescott reproduced a sketch of an apparatus using two separate receivers capable of responding to only one tone sent by one of the two transmitters. Prescott states in the text that Gray's experiments in May 1874 were important in the development of the illustrated system. The details of the transition between the May experiments and the analysing receiver are not presented. See *Electricity and the Electric Telegraph* (New York, D. Appleton, 1877), 1093–4.

65. Gray, *op. cit.* (44), 40. We have found no mention of tuning-fork experiments prior to this 1878 account, which means Gray's story, like all retrospective reconstructions, should be treated with caution.

66. *Ibid.*, 41

67. Deposition of William M. Goodridge, *Bell Telephone v. Dowd, op. cit.* (46), Part I, 86. Goodridge's account is backed by Gray in his *Experimental Researches in Telephony, op. cit.* (44), 48–9. Gray recounted that he was able to send and receive six tones simultaneously, employing his latest transmitter, which was designed to self-adjust for fluctuations in battery power, and the analysing receiver as described by Goodridge and by the contents of Gray's patent.

68. *Bell Telephone v. Dowd, op. cit.* (46), Part I, 86.

69. This intermediate spring was used on all of Gray's transmitters after he developed this improved single-tone transmitter in April 1874. See Gray, *op. cit.* (44), 23–4.

70. *Ibid.*, 40–1.

71. Bell, *op. cit.* (26), 42.

72. As Bell wrote to his parents, 5 March 1875 (Box 5, BFP): 'My lawyers were at first doubtful whether the examiners would declare an interference between me and Gray as Gray's apparatus had been there for so long a time.

'They feared I had but a poor chance—and my spirits at once fell to zero. They said it would be difficult to convince them I had not copied. When however they saw the "Autograph Telegraph" developed from the multiple telegraph —they at once said that was a good proof of independent invention as Gray had no such idea.'

73. In his 1837 telegraph, Morse did not use a transmitting key but rather a special device for automatically transmitting dots and dashes which he called his 'portrule'. This transmitter consisted of a printer's composing stick which passed underneath the right end of a long lever arm. On the left end of the lever arm was a short piece of wire that dipped into two mercury cups which were connected to a battery and the telegraph line. To send a message, Morse placed special type which had short and long ridges corresponding to the dots and dashes in his alphabet on to the composing stick. By means of a hand crank, Morse pulled the composing stick underneath the lever arm. As the composing stick moved underneath the lever, the ridges in the type caused the right end of the lever to move up and down and caused the wire on the left end to make and break contact with the mercury. By using short and long ridges in the type, Morse was thus able to use this device to send dots and dashes automatically. For further details and illustrations, see Brooke Hindle, *Invention and Emulation* (New York, W.W. Norton, 1981), 120–1.

74. A.G. Bell, 'Improvement in Transmitters and Receivers for Electric Telegraphs', US Patent 161,739 (filed 6 March 1875, granted 6 April 1875).

75. A.G. Bell, 'Telephonic Telegraph Receiver', US Patent 178,399 (filed 8 April 1876, granted 6 June 1876).

76. An account of the competition between Bell and Gray over the autograph telegraph lies beyond the scope of this paper. Bell's father, Alexander Melville, was particularly keen to have his son develop this technology. See Alexander Melville Bell to Alexander Graham Bell, 16 April 1875, Box 5, BFP.

77. Bell, *op. cit.* (74).

78. Bell learned about Gray's receiver in the course of showing his own vibratory circuit-breaker to Gray's attorney, Hayes. See A.G. Bell to G. Hubbard, 8 May 1875, Box 80, BFP.

79. See Hounshell, 'Two Paths', *op. cit.* (8), 158.

80. But Orton and Prescott, two of Gray's backers, had seen and tested Bell's multiple-harmonic-telegraph apparatus. Bell, in a letter to 'Papa and Mama' on 22 March 1875 (Box 5, BFP), reported that he had difficulty in getting it back from them after allowing them to inspect it as potential backers of his work. It is always possible that Orton and Prescott passed along useful information to Elisha Gray; our account does not reveal any gap in Gray's process that would have required this sort of information at that time. Incidentally, Orton refused to back Bell because he would enter into no 'scheme which will benefit Mr. Hubbard'.

81. Bell, *op. cit.* (26), 46. Henry encouraged Bell to pursue his telephonic researches, saying Bell had 'the germ of a great invention'. Hounshell, 'Bell and Gray,' *op. cit.* (8), discusses Bell's identification with the scientific community of his day, and includes more details on the meeting with Henry.

82. See Clarence Blake, 'The Use of the Membrana Tympani as a Phonautograph', *Boston Medical and Surgical Journal*, 1875, 92: 121-4.

83. Bell, *op. cit.* (26), 59.

84. 'Experiments Made by Alexander Graham Bell (vol. I),' notebook, Box 258, BFP, p. 13.

85. Bell described some of the earliest experiments leading to this mechanical representation in a letter to Gardiner Hubbard, 27 November 1874, Box 80, BFP.

86. Bell notebook, *op. cit.* (84), 13.

87. There is a long literature on scientific heuristics in cognitive psychology; see, for example, D. Kulkarni and H.A. Simon, 'The Processes of Scientific Discovery: The Strategies of Experimentation', *Cognitive Science*, 1989, 12: 1390-1475, and Gorman and Carlson, *op. cit.* (10).

88. Bell in this case was developing what Gruber called a 'network of enterprises', a range of related projects which facilitate each other. See H. Gruber, *Darwin on Man: A Psychological Study of Scientific Creativity* (University of Chicago Press, 1981).

89. For a discussion of the holding constant heuristic see M.E. Gorman, 'Error, falsification and scientific inference: An experimental investigation,' *Quarterly Journal of Experimental Psychology*, 1989, 41A(2): 385-412.

90. For a more detailed description of the development of Bell's liquid transmitter, see Finn, *op. cit.* (2).

91. Bell notebook, *op. cit.* (84), 38.

92. *Ibid.*, 40-1.

93. The mechanical transmitter discussed above was actually a second, improved mechanical transmitter. It was constructed from components, or mechanical representations, from the first version of this device, which was designed to produce and transmit musical tones and came from Gray's multiple-harmonic

telegraphy research. The first version employed four long reeds resting on large, toothed gears, which rotated at 1 cycle per second. Each gear had on it a number of teeth corresponding to the number of oscillations for a given musical note. For instance 128 teeth produced the musical note 'C'. These four gears were driven by a more complex gearing arrangement than that of the simpler, belt-driven, second version. See Gray, *op. cit.* (44), 25–7.

94. *Bell Telephone v. Dowd, op. cit.* (46), Part I, 124.

95. In recounting his first impressions on the transmission of speech in June 1874, he visualized a device designed to imitate human vocal chords. 'I supposed it would be necessary to construct a mechanism similar to the vocal organs of the throat, which would mould electrical waves into the same form that the air is moulded when a spoken word is uttered.' Perhaps this was Gray's initial mental model or, like Bell, Gray also began from a heuristic to follow the analogy of nature. After a conversation with a Reverend Duffield in January 1875, Gray concluded that 'it was not necessary to consider the mechanism of the vocal chords at all, but simply the physical results produced in the atmosphere by them'. See Gray, *op. cit.* (44), 36 and 54.

96. This circuit-breaker mechanism was the same mechanical representation Gray used in his analysing receiver of November/December of 1874. Bell also used a circuit-breaker for the same purposes as Gray did, namely, to sort out the individual superposed message-carrier frequencies.

97. Elisha Gray, 'Local Circuit Breakers for Electro-Harmonic Telegraphs', US Patent 194,671 (filed 15 February 1876, granted 28 August 1877).

98. Quotation is from a portion of Gray's testimony in *The Telephone Appeals. Brief for American Bell Telephone Co.*, Supreme Court of the United States, October Term, 1886 (Boston, Alfred Mudge & Son, 1887), 446.

99. *Bell Telephone v. Dowd, op. cit.* (46), Part I, 124–5.

100. *Bell Telephone v. Dowd, op. cit.* (46), Part I, 124.

101. The globe-shaped resonant cavity of the caveat receiver comes from a device that Gray constructed in December 1874. The globe resonant cavity had the same geometry as that shown in the caveat and was made out of glass. This device was tuned to respond to a single pitch so that it could be used as a receiver for a specific tone. See Gray, *op. cit.* (44), 54–5.

102. Elisha Gray, 'Instruments for Transmitting and Receiving Vocal Sounds Telegraphically, Caveat filed 14 February 1876', 1, Box 4, Fol. 4, GP.

103. *Ibid.*

104. J.E. Kingsbury, *The Telephone and Telephone Exchanges: Their Invention and Development* (Longmans, Green, 1915), 103.

105. Taylor, *op. cit.* (3), Chap. IV, p. 6.

106. On 9 December 1876, Bell filed a British Provisional Specification for 'Electric Telephony' that is even broader than his US patent; he includes the autograph telegraph, and specifically mentions the use of water as a liquid of high resistance. See *Bell Telephone v. Dowd, op. cit.* (46), Part II, 87–104.

107. Bell, *op. cit.* (26), 434.

108. *Ibid.*, 88.

109. Bell wanted to avoid interfering with an application for a spark arrester he had submitted in 1875 in which he made references to an intermittent or 'absolute break' in the current. The spark arrester employed water as a liquid of high resistance between two points in a telegraph circuit. When the Morse telegraph key interrupted the circuit, the spark arrester would dissipate the charge built up in the line so the points on the key would not be damaged by the spark that normally would occur. Bell later claimed it was this spark arrester

that gave him the idea for a liquid telephone transmitter. See Bell, *op. cit.* (26), 87.

110. See Taylor, *op. cit.* (3), Chap. XI.

111. We are grateful to Robert Bruce for first drawing our attention to this important difference. See also Kingsbury *op. cit.* (104), 106.

112. Bell, *op. cit.* (26), 85.

113. Bell experimented with a number of ways of making sound waves visible, one of which was to speak into an apparatus originally developed by Koenig that caused a flame to flicker in front of a set of mirrors. Once again, we see Bell's rich expertise with speech and sound waves playing a critical role in his invention process. See Bruce, *op. cit.* (1), 111.

114. *Ibid.*, 105.

115. Correspondence between Elisha Gray and Graham Bell, 1877, Taylor, *op. cit.* (3).

116. On 29 October 1875, Gray wrote to his attorney A.L. Hayes that 'Bell seems to be spending all his energies on the talking telegraph. While this is very interesting scientifically it has no commercial value at present, for they can do much more business over a line by methods already in use than by that system.' See Gray to Hayes, 29 October 1875, Box 2, GP.

117. Finn, *op. cit.* (2).

118. Both quotations are from Bell notebook, *op. cit.* (84), 85.

119. *Ibid.*, 97.

120. Electrical Experiments by AGB, Vol. 3', Box 258A, BFP, 32-3.

121. In his deposition in *Bell Telephone v. Dowd, op. cit.* (46), Part I, 142-3, Gray suggested why he dismissed the magneto design: 'I thought it would be impossible to make a practical working speaking telephone on the principle shown by Professor Bell, to wit: generating electric currents with the power of the voice, as it seemed to me then that the vibrations were so slight in amplitude and the inductor necessarily so light that the currents thus generated would be too feeble for practical purposes.'

122. See Gorman and Carlson, *op. cit.* (33), 146.

123. Taylor, *op. cit.* (3), Chap. IX, 10.

124. In terms of the framework, this pattern is an inferred goal because Gray never explicitly stated that he would put his future efforts into blocking the Bell patent, but his actions point to those intentions.

125. Elisha Gray, 'Art of Transmitting Vocal Sounds Telegraphically', US Patent Art application filed 29 October 1877 and 'Improvements in Electric Telephony', US Patent application filed 29 October 1877, in *Bell Telephone v. Dowd, op. cit.* (46), Part II.

126. Elisha Gray, 'Speaking-Telephone', US Patent 204,029 (filed 21 March 1878, granted 21 May 1878).

127. D. Klahr and K. Dunbar, 'Dual Space Search during Scientific Reasoning,' *Cognitive Science*, 1988, 12: 1-48.

128. As Bell said in a letter to his parents on 23 November 1874 (Box 4, BFP), Gray 'has the advantage over me in being a practical electrician—but I have reason to believe I am better acquainted with the phenomenon of sound . . .'

129. Gray summarized his style in the following way: 'There is a vast difference between doing a thing in the laboratory and on the line where it is to be used. Many instruments and systems work splendidly on short circuits that utterly fail on long ones. A long circuit has several additional conditions to be taken care of besides mere resistance. My long and large experience in matters pertaining to telegraphy enables me to work understandingly. I therefore reject at a glance many things that would seem to one unacquainted with the freaks of long line phenomena something very valuable.' Gray to Hayes, May 7 1875, Box 2, GP.

On the Origins of the Suction Lift Pump

GRAHAM HOLLISTER-SHORT

One of the more notable mechanical inventions to come out of Hellenistic Egypt was the force pump. This was supposedly the creation of Ctesibius of Alexandria (*fl.* 250 BC) but first described by Philon of Byzantium only some one hundred years later. Philon's description of the Ctesibian pump, and later the remains of examples recovered from Roman sites, show that it was usual to place two such pumps side by side. If both pumps discharged into a common rising pipe, as seems always to have been the case, then the two pumps, working alternately, would have been capable of forcing upwards a continuous stream of water.[1] Since in antiquity, and indeed until the fifteenth century in Europe, no attempt appears to have been made to apply more than manual labour to the working of such pumps; any such ensemble was necessarily small in size and of very limited capacity. But small is beautiful—or was in this case. The mechanical arrangements of the ancient force pump, in so far as they are revealed by surviving remnants, display an impressive capacity for machine design.[2]

In describing the force pump's mode of operation it is a good deal easier to tackle one such unit than two, and so one it will be (Figure 1). The barrel, or cylinder, of the pump was fitted with two valves. One, the foot valve, was let into the base of the cylinder, the other into the lower part of the cylinder wall. These valves opened and shut alternately, each responding to the motion of the piston. The piston, a valveless, *solid* plug, was kept

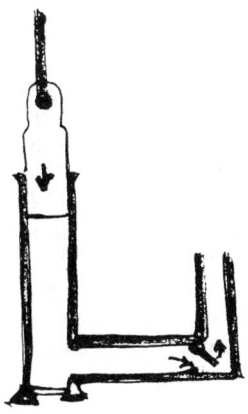

Figure 1 Force pump of Ctesibian type: down-stroke.

57

in airtight contact with the cylinder walls by means of a leather covering. As the piston was drawn upward a 'suction' effect was produced in the space created beneath it, causing the foot valve, that could only open inwards, to rise. Since the ancient force pump cylinder was not fitted with a supply pipe leading down into the water supply, its base had necessarily to stand in the water itself. As the piston moved upwards, therefore, atmospheric pressure, acting on the external water surface, forced water directly into the cylinder through the foot valve. The same drop in pressure that forced open the foot valve caused the side-wall valve, only able to open outwards, to remain tightly shut. As the piston performed its downward excursion the valve positions were reversed. The water in the cylinder, under pressure from the descending piston, forced the foot valve shut, while the same pressure forced open the side-wall outlet valve into the rising pipe through which was expelled the charge of water in the cylinder. Successive increments of water, forced into the rising pipe by the piston's successive downward excursions, caused the water column in the pipe to build up until it reached the point of discharge. Obvious though it may be, I must all the same insist on noting here that the water, both on entering and on leaving the pump, was an entirely passive entity throughout the operation. It was at all times being acted upon by other forces: first by atmospheric pressure and then by pressure exerted on it by the piston's downward travel.

In the ancient world such pumps supplied water for domestic purposes, and may have acted also as fire-fighting equipment. A new application arose with Callinicus's invention, *c.* AD 765, of Greek fire. Little information has survived as to the construction of the Greek force-pump flame throwers in use from that time. However, a Chinese description of 1044, the only surviving description of a Greek fire flame-thrower pump according to Joseph Needham, allows us to suppose that in the West, as in China, suction or feed pipes supplying the force pumps from the reservoir, although perhaps of no great length, had long been in use in such devices.[3] When the suction lift pump appears some seven centuries later, it too will have this additional member.

So much for the history of the force pump. The history, and in particular the origin, of the suction lift pump is even more heavily veiled. It is in fact, with the possible exception of the crank, one of the most obscure problems in the history of technology. The first evidence of its very existence comes in a sketch of about 1425. It occurs in a collection of drawings of machines and devices composed by Mariano Taccola (1382–post-1453) in the period 1419 to 1433, and gathered finally under the title *De Ingeneis* (On machines).[4] Taccola was well on the way to completing the first two books of the four comprising this collection by 1427, so that the drawing, on an early folio in Book 2, may perhaps be thought of as having been done already by about 1425.[5] It is not, of course, the only drawing of water-raising machines to be found in Taccola's collection. Some of them are, in my opinion, of equal importance with the suction lift pump itself. My object in this paper is to take the drawing of 1425 and certain of these other pumps and, by viewing them as a group, try to understand the context in which the invention may have been conceived, and, by the same token, open up the possibility of explaining how this new kind of pump may have come into existence.

Before I describe the suction lift pump's mode of action, or even offer a definition of the term, some preliminary points need to be made. This done, it will be clear how radically different in character the ancient force pump and medieval suction lift pump really were, one from the other. This difference, one is obliged to say, will appear not so much perhaps in respect of their physical form as in the way in which work was done in each. Indeed, this last is the very thing which in my opinion makes it unlikely that the second could, in any very meaningful sense, be said to have derived from the first. Even in respect of their physical form, however, one possibility that might be urged (and which I shall in fact urge) is that on this very point the similarity of the two pumps might just as easily be the outcome of some process of convergence as of some significant degree of affiliation. As for affiliation, the fact is that no one has suggested a plausible conceptual route that might lead from the solid valve of the one to the pierced valve of the other. The conventional view of the matter, in so far as one can be said to exist, seems nevertheless to be that the suction lift pump derives from the force pump of antiquity. Such a view rests on no better basis, it seems to me, than that of *post hoc ergo propter hoc.*[6]

The essential features of the suction lift pump and of its mode of operation may now be briefly stated (Figure 2). The suction element in the name refers to the well-known fact that water will rise up a pipe from which air has been exhausted to a maximum of about 32 feet, this maximum decreasing with elevation above sea level. Atmospheric pressure, acting upon the external surface of the water being pumped from, is, of course, responsible for the phenomenon, although it is extremely unlikely that any of the craftsmen who began establishing this limit as an experimental fact in the process of creating the pumps to be discussed here had any idea as to its cause or the reasons for its limited operation.[7] In any event, the suction lift pump possessed a suction or supply pipe, of variable length, reaching down into the water to be lifted. Where this pipe joined the cylinder base a foot valve, opening only upwards, was fitted. This opened when the piston rose but was forced shut as it moved down the cylinder. The piston here, unlike the solid plug of the force pump, was *pierced*, its aperture fitted with a one-way valve. The piston valve, working in an opposed sense to the foot valve, was shut by atmospheric pressure on the piston's up-stroke but forced open on the down-stroke by the pressure of the water against which the piston was acting. This water, of course, passed through the valve to lie on the back or the top side of the piston. Successive strokes and successive increments of water built up to form a water column riding up and down on the back of the piston as the piston reciprocated. This constituted the lifting aspect of the suction lift pump. The column would build within the cylinder walls until it reached the point of discharge and then break off, as it were. The presence of this valve marks one great point of difference between the two types of pump. Another, equally critical difference was that the valved piston eliminated all need for a separate rising pipe. Two points are of some importance here. First, it will be clear that

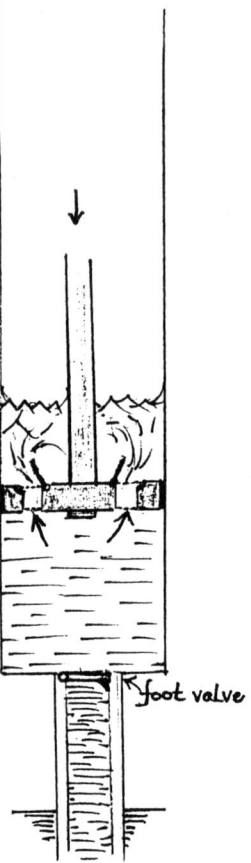

Figure 2 Suction lift pump: down-stroke.

the proportioning of suction to lift could be adjusted almost at will. The
upper limit on how much suction was to be employed was plainly set by
nature, but in practice the pump masters, for so they should be called,
would soon have learnt that it was unwise to exploit suction to its limit.
The penalty for doing so was an economic one: a higher rate of wear and
damage to the valves. According to one of the most acute seventeenth-
century writers on mining machinery it was to be noted that 'the deeper
you let fall your bucket, the easier it will go, and not so soon be subject
to be out of order; it is better to lift a foot of water than suck or draw a
foot'.[8] The bulk of the work was generally done by lift, not suction. Hence
the undesirability of statements one encounters so frequently in the
literature that such pumps were of limited value since they could only suck
water to a maximum of about 30 feet.[9] The combination of suction and lift
knew no such constraint. As to lift, nature was not so niggardly. How much
lift could be obtained depended on the growing skill of the craftsmen in

fabricating pump barrels. And, if this was not in question, then the limit was set ultimately by the capacity of the materials used in fabrication to resist pressure, and, even more importantly, by their cost.[10]

The second point is pre-eminently worthy of consideration and will be appealed to repeatedly in the discussion which follows. In a suction lift pump the water column, as it continually rode up and down on top of the piston, was really pumping itself. This is immediately apparent if one thinks of it as acting exactly as if it were a solid ram, but one whose shape was imposed upon it by the cylinder-pipe walls, while for the necessary hardening of its 'head' it had the solid cap provided by the piston 'bucket' or piston disc. On this disc the whole weight of the ram reposed. The water itself was therefore an active component of the pump ensemble, in contrast to its inert character, as so much dead weight, in the rising pipe of a force pump. This active role that some nameless inventor imagined for the very water that was to be lifted is, in my opinion, a conceptual stroke of such brilliance that the invention of the valved piston with which of course it is inseparably linked, and on which attention is ordinarily focused, must appear by contrast as a secondary feature; no more, in fact, than the means by which this other, greater, effect was to be secured. The context in which this invention is to be situated is no less interesting. The exploitation of one natural force, atmospheric pressure, was old in pump work. Exploitation of gravity, however, as in the free fall of the water column in the suction lift pump, was new. But then the exploitation of gravity was, in peace as in war, something of a medieval triumph: the weight-driven clock, the counter-weighted trebuchet, and now the suction lift pump, the latter, it might be said, completing not unworthily this trinity of ingenious devices.

The radically different nature of these two species of pump—the force pump and the suction lift pump—will now be clear. As has been mentioned already, it has proved a matter of extreme difficulty to propose a convincing conceptual route from the unpierced piston of the Greeks to the active water column and valved piston of the late medieval period. The literature is virtually silent on the matter. The difficulty, as one can see, arises in part from the fact that evidence of sufficiently early date that might throw light on the genesis of the suction lift pump has not so far been discovered. It is only with the work of Mariano Taccola in the 1420s that the first drawings become available. It is more than likely that they are all considerably after the fact, and that the invention of the suction lift pump had probably been made some considerable time before Taccola put pen to paper. Late in the day though his drawings probably are, they do nevertheless provide a number of backward-pointing clues to what may have been involved in the invention. Enquiry having therefore to begin where it must, one turns to the group of pump drawings already referred to that were executed in the period before 1433 by Taccola, the so-called Archimedes of Siena. All the drawings come from four books of machines that he gathered under the title *De Ingeneis* (below D.I.) and which he completed on 13 January 1433. The four books were given rather fetching titles: Book One is called the first book of the lion, *Liber primus leonis*, the second, the book of the dragon, that is, the *Liber secundus draconis*. Book Three concerned machines and structures

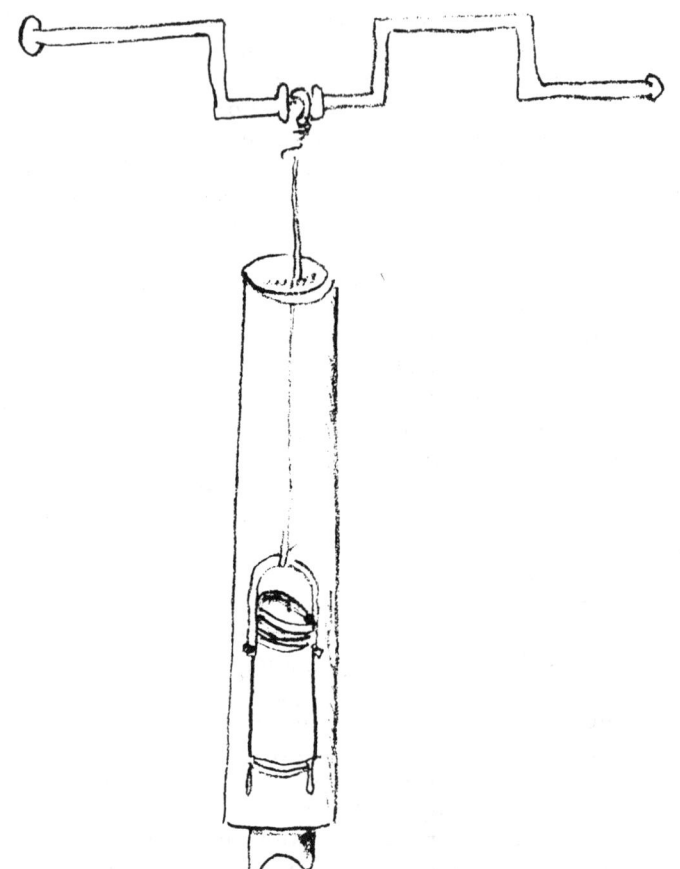

Figure 3 Taccola: *De Ingeneis* II, f82v: suction lift pump, *c.* 1425.

not yet in common use, that is, the *Liber tertius de ingeneis ac edifitiis non usitatis*, and the fourth was about everyday structures, that is, the *Liber quartus de edifitiis cotidianis*. Some of these drawings are repeated in Taccola's second great work, the ten books of machines, *De machinis libri decem* (below D.M.), completed in 1449.[11] Four of Taccola's drawings call for comment and analysis. They are, in the order in which they appear in his work: (i) the suction lift pump proper, D.I. I 53r , D.I. II 82v and D.I. III 29r (Figure 3); (ii) the sack pump, D.I. II 85r and D.M. 40r (Figure 4); (iii) the solid ram pump with cannelure, D.I. III 34r and D.M. 39r (Figure 5); and (iv) the suction lift hollow tube ram pump, D.I. III 45r (Figure 6).

I shall begin with (iv), which, in the absence of an authentic terminology, I have had to burden with a rather cumbrous name. As will be clear from Figure 6, it consisted of four hollow, mobile tubes worked in pairs, off a manually powered camshaft. If Taccola's drawing was done to scale, then the tubes were about 10 feet long. He specified that the tubes were to be

Figure 4 Taccola: *De Machinis*, f40r: sack pump, 1449.

of tin, their bases fitting closely into the dwarf fixed tin cylinders in which they reciprocated. Whether the bases of the dwarf cylinders were submerged in water or, standing above it, received their water through supply pipes is not entirely clear. The action of the pump was in any case hardly affected, and was the same as that of the suction lift pump proper described above. Although the tubes worked in pairs, it will be convenient to describe the action of one tube only. As the cam began to lift the hollow tube, really nothing more than an elongated piston or bucket, so atmospheric pressure forced water into the cylinder through the one-way valve in its base and shut the one-way valve in the bottom of the tube. The action of the valves was reversed when the cam, releasing the tube, allowed it to fall back and take in the water at the bottom of the cylinder. As the tube received further pulses of water and filled up, so it more and more approached the condition of a 'solid' ram. In this respect it displays the essential nature of the suction lift pump but certainly not its fully rationalized form. A hollow tube to contain a moving water column is one thing, but a *mobile* hollow tube is, in terms of the suction lift pump proper, a totally irrational feature. In respect of the piston, only water needed to move for the job to be done. Seen in this light (iv) may be regarded as an incompletely realized, and therefore perhaps an early form of, suction lift pump. The mobile hollow tubes in (iv) were redundant. What makes them interesting, however, is that they have every appearance of being skeuomorphic. The phenomenon

Figure 5 Taccola: *De Ingeneis* III, f34r: solid ram pump with cannelure, *c.* 1433.

of skeuomorphism is well known to students of primitive arts and crafts. Less well understood is that the phenomenon is by no means confined to primitive peoples and their crafts. It is a constant in human psychology, as far as one can see. Skeuomorphism refers to the well-known fact that a newly devised artefact commonly retains features that are, strictly speaking, quite redundant. As an example one might take a cast-metal socketed spearhead. Such a spearhead may well display a pattern of lashing marks. This pattern is a simulacrum of the pattern that real sinew lashing formed in the days when metal was not known and only flint was available as spearhead material. Only by lashing it to the shaft could it be kept in place. Obviously a socketed metal spearhead required no such lashing, and yet such was the organizing power of the gestalt form, or image, of the spear on the imagination of the craftsmen that, for a time at least, they were quite unable to break clear of its spell. Skeuomorphs therefore are of inestimable value in seeking to retrace the path along which an invention has travelled. They are, one might say, something like spoor marks. They necessarily lead

Figure 6 Taccola: *De Ingeneis* III, f34r: hollow tube ram pump, *c.* 1433.

back in time to the material form of some earlier device from which the new
entity has been derived: the skeuomorphic features, as it were, carried
across from the old situation to the new.[12] In other words, it is as if the
energy of the inventive act expends itself on an existing form or gestalt and
in the process renders it plastic enough to allow the function of the newly
conceived device to be released from its 'host'. Whatever does not
materially frustrate that function does not in principle need to be altered,
and in all probability will not be altered, at least to begin with. So it is that
a new device almost always enters the world with shreds of old clothing
hanging about it. The process of critical revision, by stripping them away
(which amounts also to a very thorough effacing of evidences of origin), is
also, to put aside metaphor, the psychological recognition of the new entity
as a new gestalt endowed with its own logical or organizational coherence.

It is at this stage, and only at this stage, that redundant features can actually be seen to be redundant.[13] Viewed in this way the mobile hollow tube might seem to indicate descent from some ancestral device in which the working ram, as in pile-driving, was actually solid, performing its percussive work by gravity on a solid object. At all events water was now in effect pile-driving itself.

At this point it is useful to introduce (i), Taccola's suction lift pump proper (Figure 3), into the discussion. The recognition of what was necessary to achieve a full realization of suction lift pump form is here much more nearly achieved than in (iv). The mobile pipe has gone; only the piston moves. Its down-stroke will, as before, be performed under gravity, as the rope attaching it to the crankshaft above indicates. It is, however, a peculiar piston, even though it is no longer the grossly elongated piston or mobile tube of (iv). The piston or bucket of (i) might appear to be a truncated form of the mobile tube. If this is so, then this truncated form is itself still skeuomorphic since a piston in its fully rationalized form needed to be not a quarter or length of the cylinder (as here) in which it reciprocated but simply 'a disc of iron one finger thick'. This was how Agricola explained matters in the 1540s as he drafted Book 6 of his *De Re Metallica*.[14]

Before I discuss the solid ram pump with cannelure of Figure 5, or even explain how it is that the word 'cannelure' comes to be here at all, something needs to be said about the history of one of Taccola's manuscripts. What is now known as Codex Latinus 197, in the possession of the Bayerische Staatsbibliothek, once belonged to Johann Widmanstetter. He acquired it in Siena some time in the period 1533–42, and subsequently returned to Germany with it. Before that the manuscript had previously belonged to the great Sienese architect and engineer of the generation after Taccola, that is, Francesco di Giorgio Martini (1439–1501), whose own manuscript, the *Trattati di Architettura, Ingegneria e Arte Militare*, was completed about 1475.[15] Perhaps it is no surprise then to find many of Taccola's machine sketches reproduced in Francesco's pages. There they are rendered in a style of great brilliance, often superior in its attention to detail to even Leonardo's machine sketches. Among the machines taken from Taccola one may find, in several configurations, the solid ram pump with cannelure. From these I have selected the one closest to the mechanical arrangements of the Taccolan original.[16] Because Francesco di Giorgio used very freely the depictional device of portraying his machines as if they were made of transparent material, clues as to the construction of Taccola's pump, which are lacking in Figure 5, are now to be seen very clearly in Francesco's drawing of the same machine (Figure 7), notably the deep grooves or cannelures cut into the bodies of the ram pumps. Armed with this information it is possible to go to work on (iii).

The pair of ram pumps worked by swape levers in Taccola's drawing are comparable in size with the mobile tube of (iv) since he speaks of them as being 'potest aqua elevari a *d* usque *e* per quinque brachii' ('able to raise water from *d* to *e* through five "arms" '); i.e. about 10 feet. The rams were cylinders of wood and were without valves. They had to be able to drop 'usque ad fundum cannarum' ('to the very bottom of the tubes') they

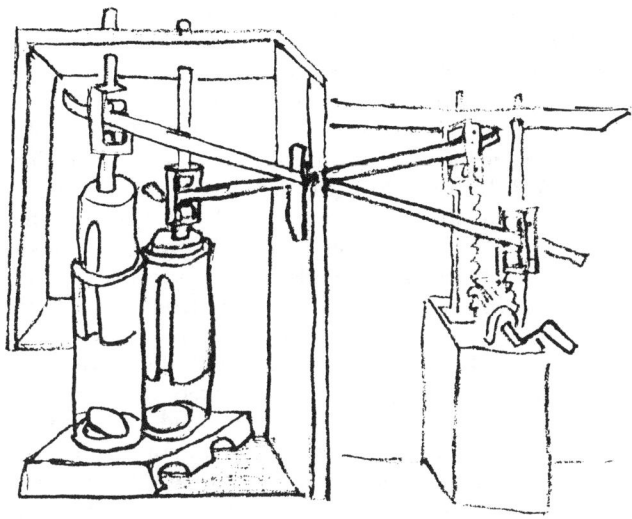

Figure 7 Francesco di Giorgio Martini, *Trattati* . . ., f45v: solid ram pumps with cannelures, *c.* 1475.

moved in.[17] It was indeed a *sine qua non* of the ram pump that each ram should in fact fall to the very bottom of its tube, since otherwise it could scarcely have worked at all. This was because the ram in each case, for perhaps about 60 per cent of its length (beginning at the bottom), had a groove cut into it. This was for the passage of the water that was to issue from the spout, marked e in Figure 5, at the top of the tube. Now let us suppose that the ram is being winched up its tube. For the first foot or two of its upward travel, however, only the solid, uncut upper portion of the ram will be sliding past the spout hole. During this phase the ram would be producing a vacuum in the tube beneath it, thus causing atmospheric pressure to force water into the tube through the one-way valve (animella) in the little supply tube (cannicula) set below the base of the tube. So far the ram was performing like any regular piston, although only for as long as it took for its grooved portion to rise into alignment with the spout hole. Obviously this at once cut off any further vacuum effect and caused the foot valve to shut. When the ram was allowed to drop back under gravity, the water beneath would be violently propelled up the groove and voided before the solid upper portion of the ram again closed off the spout hole.[18]

One might easily conclude that as a pump it was more primitive in form, and certainly likely to be less effective in action, than (iv) and much less so than (i), and, following this, that chronologically it might have come into existence before either. The danger in labelling the solid ram as primitive, however, is to appear to slight, if not to write off altogether, the effort of mind which had gone into its elaboration. For the fact of the matter is that with the sole exception of the slide valve figuring in Tsêng Kung-Liang's *Collection of the Most Important Military Techniques* of 1044, it is, unless the idea had by then or later travelled to the West, the first slide valve of this type to appear in Western technology.[19] It will not be repeated until 1873, the year in which John Darlington demonstrated his valveless compressed-air drill.[20] The ram, therefore, obviously calls for further consideration. Might it not be, in fact, the very intermediate form (or one of several, for all we know) between force pump and suction lift pump which has so far eluded researchers? This would be a nice irony indeed, since I have cast doubt on the idea that there can be any real link between the two kinds of pump at all. Prima facie, the ram pump does appear to possess features common to both. In force-pump mode the transport of water is achieved by force exerted downwards and it is certainly true that the ram lacks any kind of conventional valved piston. It is, however, when one looks at features shared not with the force pump but with the suction lift pump that one recognizes where real affiliation is more likely to reside. As in the one, so in the other, the down-stroke was performed by gravity, a decisive pile-driving kind of similarity, it seems to me, and one sufficient on its own, in my opinion, to render the possibility of any force-pump ancestry unlikely. Secondly, and in a very limited fashion, the water and ram of the ram pump constituted one body in the sense that they were contained together within a common cylinder. Comparing it with the suction lift pump in this respect would certainly be putting the evidence under unwarranted strain, unless there were something further to be brought in support of the idea. What sanctions it, I think, is the feature of quite extraordinary subtlety mentioned above. In the ram pump's operation, the ram itself might have no valve but it is undeniable that ram and tube in combination constituted a slide valve system. How else could one describe the precisely controlled opening and shutting of the spout hole (the exhaust port) in the cylinder as the grooved and ungrooved portion of the ram moved past it, and on which the action of the pump depended?

It might of course be argued that Taccola's drawing, even though it was copied by Francesco di Giorgio, was merely a paper scheme, an interesting idea. All kinds of questions could be asked about it: how deep was the groove in the ram, and how this groove was to be proportioned to the ram diameter, and suchlike matters. On all these points Taccola is silent. No doubt secrecy was some defence against plagiarists, but no doubt, also, these and many other matters were to be determined on site according to rules known to the initiated. On one other point, however, since Francesco di Giorgio's drawing has been used to throw light on Taccola's earlier sketch, we might ask how Taccola proposed to maintain an airtight seal between the uncut portion of the ram and its tube. It will be obvious that as the swape beam winds up the ram the circular movement of the end of the swape beam will have the effect of forcing the ram to one side of its cylinder. No doubt the loose jointing of swape beam to ram top would deal to an extent with this problem, but not altogether. This is why it is interesting to note in Francesco's drawing that the problem is addressed by means of anti-friction rollers, through which the swape beam can slide without wrenching the piston rod attached to the top of the ram out of its proper straight-line motion. Of course there is a certain danger in taking these sketches too literally. They were meant to be vehicles for ideas, not guides to construction. In Taccola's own drawing in Figure 5 the spout nearer to the viewer can be seen ejecting a tremendous jet of water. It will also be observed that the ram in this particular tube is actually being drawn upwards, so that any ejection of water would be impossible for it in this position. Francesco's two grooved rams in Figure 7 are really no better guide either. The left-hand ram, for instance, has twice as much uncut head as the right-hand one, and it is doubtful if either could have worked as shown. Many of Leonardo's sketches, for that matter, are plainly unworkable and have to be understood for what they are, vehicles for ideas.[21]

However, to speak of ideas in the context of Figure 5 raises a further question which has to be addressed: was Taccola himself the inventor of the ram pump? In an early folio of Book 1 of his *De Ingeneis* Taccola claims four inventions as his own. Nor is this all, for he states that each had been 'tested' in 1427. The third of these inventions relates to f33v and 34r of D.I. III, his double-page drawing showing, of course, alongside other features, precisely the grooved ram pumps in question here. But all is not plain sailing, as will appear from his claim 'that grist mills can be constructed on legs, without incline, in Gallic or country style as desired by the miller'.[22] With nothing more than this to go on it is rather difficult to know precisely what Taccola was claiming as his invention. Prager and Scaglia, his editors, believed that what he meant to claim priority for here was the idea 'of a pump actuated recirculation from the lake (shown here as a pond) into an elevated tank'.[23] To put the accent on 'without incline', i.e. with no outfalls, as they implicitly do here, and therefore pinpoint *recirculation* seems not unreasonable, although in taking matters no further they scarcely do justice either to Taccola's claim or to their own argument.

On the very next folio, 34v and 35r, Taccola draws a tide mill in which recirculation of water was achieved by natural means, using the flux and reflux of the sea. If this juxtaposition is thought unlikely to have been merely fortuitous, then may not Taccola have been seeking to make the point that *ingenium*, by supplementing nature where nature was deficient, could, so to speak, complete nature?[24] This is a big enough thought. It is also examplary of the kind of thinking that it would be natural to expect in a technologically dynamic culture, such as by common consent was already well developed in the West at this time. It is the groundswell informing Leonardo's work, and later on the Machine Books also. On the level of practice, recirculation was to be an important feature of Western technological development. The work of returning the tail-race water of a mill back to the storage reservoir which fed it was, for example, a task given very early on to the steam engine. James Watt's first machine was so employed at Soho, but the idea was already old then, as we have seen.

Whether or not Taccola was the inventor of the ram pump, although it might seem on balance that he was not, that pump was still an extraordinary stroke. It was a creation that the dialectical masters could have admired: for as between the disjunctive, either/or, of force or suction lift, here, against all the odds, was a *tertium quid*! With Taccola dismissed from the story, except in the role of transmitter, it is still difficult to see where the ram pump should be placed in the evolutionary sequence. Was it ancestral to the suction lift pump proper, and therefore earlier than the hollow-tube pump as well? If so, how many stages might there have been before a solid ram with groove would finally acquire (i) a central bore hole, (ii) a bore hole taken through the entire length of the ram, and (iii) a valve of conventional type closing the bore? Or might there have been a one-step saltation straight from ram pump to the hollow tubes? Whatever the answer might be, there is no doubting one thing, that here we see a capacity for sophisticated analysis and design of considerable amplitude. This serves notice on us that at the highest levels of the hydraulic engineering fraternity, whether it was about 1400 or 1350 or earlier, one should expect to find talents equal to this further task. There would be others no less capable than the anonymous inventor or inventors of the three pumps discussed here of sustaining and augmenting in their turn the experiential tradition of which they were all—in the fullest sense of the word—carriers. In this way, the mechanical inventories these men commanded were constantly improved and then generalized throughout the cultural and technical ecumene of western Europe.

It will not have escaped notice that nothing has yet been said about (ii), the sack pump (Figure 4). The sack pump was in effect a suction lift pump. The sack, like the kind used for straining wine, Taccola advises us, and presumably filled with something to bulk it out, was lowered into the cylinder and sank through the water to the bottom. On being sharply raised the vacuum effect below and the weight of water above the sack caused it to deform, thereby producing a seal between the sack and the cylinder walls. This upward movement, of course, caused the foot valve to open and allow a fresh supply of water to enter the cylinder. In this way water was lifted on the back of the sack to the point of discharge, or, since this would be wasteful of energy, more probably the sack was worked up and down until the growing water column moving with it mounted to that point. The sack in either case was performing exactly as if it were a valved piston. The improved sack, in the rationalized form of a conical leather diaphragm, survived on its own account and is described and figured in Agricola's work.[25] Ewbank describes it in some detail under the title of bilge or burr pump. The sucker of the bilge pump still in the nineteenth century consisted of a hollow cone or truncated cone of strong leather. Ewbank rather nicely describes the action of this sucker as 'something like moving a parasol up and down in water; the sides close as the rod descends and open when it rises. It is the simplest modification of the sucker known and probably the most ancient.'[26] This was Ewbank's opinion and he may well be right. But it seems to me that without knowing how old (or how new) the device may have been when Taccola sketched it, one can only guess at where it might belong in the line of development. The somewhat lyrical note that Taccola sounds in describing this pump, 'hoc attingere est pulcherimum' ('this way of lifting water is the most beautiful'), might suggest that it was still enough of a novelty in his time to compel admiration.[27] My own guess is that it is later than the hollow tube ram pump, possibly even later than the

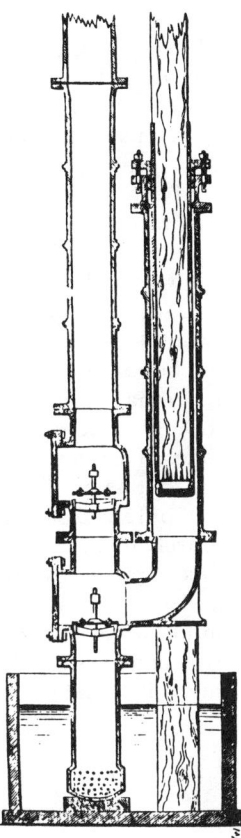

Figure 8 Cornish system pumping unit *c.* 1820 with solid plunger and rising pipe: J. Farey, *A Treatise on the steam engine* . . ., 1971, Vol. 2, p. 142.

suction lift pump proper. In that case, Taccola's pleasure might have been elicited by the unexpected sight of something as simple as a sack doing the work of a much elaborately contrived 'bucket'.[28] We also observe what we might well have thought impossible in such a small sample of pumps as have been surveyed in this paper: nothing less than a second *tertium quid*, making mock of the simple verities of pistons, solid or pierced.

Through all these suction lift pump variations the organizing idea conserves a continuously recognizable identity. When, later, the suction lift pump proper is developed as a heavy-duty machine element in mine pumping, this identity appears in ever sharper focus. By about 1800 a number of water columns, reciprocating with their pump pistons and arranged serially in deep and heavily watered mines, might each easily exceed 300 feet. During the 1790s, however, what is known as the Cornish system of pumping began to displace traditional practice. Dickinson characterized the new method as one in which the engine pumped the rods and the rods pumped the water, but until one has a clear picture of what the hardware looked like this really very economical description is unlikely to be helpful. In the new configuration (Figure 8), the water no longer rose and fell with the piston rod as the rod reciprocated in the pump barrel (perhaps some 300 feet long, let us remember), but was forced into a separate rising pipe by the solid piston of a plunger pump. Such pistons, replacing the valved type, acted exactly like the force pump of Figure 1. Does it not seem, on the face of things at least, that a full circle had been turned back to the days of Ctesibius? But with these pumps nothing is quite so simple. The Cornish system was still a gravity system. In that system the shaft rod, perhaps 20 inches square and a thousand or more feet long, once it had been lifted to the top of its stroke by the engine, was released and then went into virtual free fall on its down stroke. Was this not pile-driving to end all pile-driving, so that even as the new system replaces the old the essential nature of the old system is conserved? To understand events in this way is to place oneself at no great distance from Simondon's view of technological evolution. Put as briefly as possible this is that as an invention moves from a phase of purely abstract being, *the idea of the machine* in the mind of the inventor, a purely mental construct, and acquires a three-dimensional reality, so it begins to move towards progressive saturation.[29] This has some slight correspondence with Usher's idea of critical revision, but with this difference: Simondon would insist that the developing internal logic of the technical object will increasingly determine the direction that revision will take. The last state will reveal the essential idea, or *the machine as idea*.

Endnote

In a copy of the *De Machinis* made about 1470 for Paolo Santini (Codex Parisinus Latinus 7239) a drawing of a further type of pump occurs (at f39v) not to be found in the original MS of 1449 (Codex Latinus Monacensis 28800). This shows a bellows pump. The pump consists of two cuneate concertina leather bags, each with a suction pipe reaching down to the reservoir. The bags are depressed and opened alternately by a man transferring his weight to one while pulling open the other. Since, however, water was raised only by suction in this device and was exhausted with no lifting or forcing stage, I have excluded pumps of this kind from the discussion. They are modelled very evidently on the customary form of metallurgical bellows. For a discussion of the MS of 1470 see: Mariano Taccola, *De Rebus Militaribus (De Machinis 1449)*, ed. and trans. E. Knobloch (Baden-Baden, 1984). For the drawing and text of the bellows pump see pp. 208-9.

Notes and References

1. A twin pump unit was unavoidable if an uninterrupted flow was required. The double-acting force pump is not attested in the West before the 1470s.

2. For drawings and descriptions of two Roman force pumps from Bolsena see British Museum, *Guide to Exhibition Illustrating Greek and Roman Life* (London, 1920), Figs. 127, 128 and pp. 120-1. The figures are reproduced in S. Shapiro,

'The origin of the suction pump', *Technology and Culture*, 1964, 5: 566–74. This last, despite the promise of its title, has nothing to say on the problem of origins at all.

3. J. Needham, *Science and Civilization in China*, Vol. 4, Part 2 (Cambridge, 1965), 144–9 gives the fullest account, but see also Vol. 5 Part 7 (Cambridge, 1986), 81–94 for a partial recapitulation. Tsêng Kung-Liang was the editor of the 1510 edition of the Wu Ching Tsung Yao, compiled by imperial order in 1040–44. It is the oldest extant version of the text.

4. F.D. Prager and G. Scaglia, *Mariano Taccola and His Book 'De Ingeneis'* (Cambridge (Mass.) and London, 1972), edit and transcribe selected texts with their accompanying drawings from each of the four books. The complete text has not been published. J.H. Beck, *Mariano di Jacopo detto il Taccola, Liber tertius de ingeneis de edifitiis non usitatis* (Milan, 1969), has edited and transcribed a facsimile reproduction of Book 3 of the *De Ingeneis*. In neither case, however, is any very serious analysis attempted of Taccola's machines and devices.

5. Prager and Scaglia, *op. cit.* (4), 34–5.

6. The only recent paper on the suction lift pump is that by S. Shapiro, cited above (2), in which this idea implicitly shapes the discussion.

7. The role of atmospheric pressure in the action of the suction lift pump was first explained by Isaac Beekman in 1618. On this and other matters see C. de Waard, *L'expérience barométrique . . .* (Thouars, 1936), *passim*. It is, however, Evangelista Torricelli's resonant phrase of 1643, 'noi viviamo nel fondo d'un pelago d'aria elementare', which is remembered in this connection. On Torricelli see W.E.K. Middleton, *History of the Barometer* (Baltimore, 1965).

8. R. d'Acres, *The Art of Water Drawing* (London, 1659), 34, reproduced in facsimile and introduced by Rhys Jenkins as Newcomen Society, extra publication No. 2 (Cambridge, 1930). The pseudonym d'Acres is thought to have concealed Robert Thornton (1618–79).

9. S. Shapiro, *op. cit.* (2), 568. In his classic description of mining engineering, the *Anleitung zu der Bergbaukunst* of 1773, 321, Christoph Delius noted that 'Die Saugröhren werden bey den hiesigen Kunstsätze, nicht über zwey Klafter gemacht, und die stehen bey zwey Schuh tief im Sumpfe. Da nun der Hub bey den hiesigen maschinen gemeinlich 6 Schuh hoch ist, so darf die Wassersäule in dem luftleeren Raume bis unter die hochste Erhohung des Kolben nicht über 16 Schuh stiegen.' ('Suction pipes in the pumps here [i.e. Schemnitz] are not made more than 12 feet long and stand two feet deep in the sump. Now given that the piston stroke in the machines here is usually 6 feet, it follows that the water column in the exhausted space, even when the piston is at the limit of its upward excursion, never climbs higher than 16 feet.')

10. Something of a debate raged among seventeenth- and eighteenth-century German mining engineers as to whether 'hohe Sätze' (high pumps) sucking about 15–20 feet and lifting 70 feet, or 'niedrige Sätze' (low pumps) sucking about 25 feet and lifting 5 feet, constituted best practice. C.T. Delius, *op. cit.* (9) (Vienna, 1773; reprinted Prague, 1975), 332, had the last (and favourable) word. In Agricola's description of 1556 suction and lift are equally proportioned.

11. Scaglia, *Mariano Taccola: De Machinis*, 2 vols. (Wiesbaden, 1971). Volume 1 presents the Latin text and English translation, Vol. 2 is a facsimile reproduction of the ten books.

12. R.U. Sayce, *Primitive Arts and Crafts: An Introduction to the Study of Material Culture* (Cambridge, 1933; reprinted New York, 1963), 84: 'Not only is the general shape of the earlier object copied in the new materials, but even in matters of detail and ornamentation many objects throw light on their own genealogy.' A modern example might be the way the design of direction-changing indicators on cars has

evolved (in Britain at least) from 'trafficators', arms with lights, springing out of the side panels of the car, to flashing indicator lights fore and aft, their modern rationalized form.

13. 'Seeing and seeing as' is N.R. Hanson's very concise description of the before-and-after effect of a gestalt switch. See his *Patterns of Discovery* (London, 1962), 5, on Kepler and Brahe.

14. G. Agricola, *op. cit.* (10) (Basle, 1556), 134; 'orbiculus ferreus digitum crassus'. Wooden pistons, of course, needed to be five or six times thicker than this.

15. C. and L.M. Maltese, *Francesco di Giorgio Martini Trattati . . .*, 2 vols. (Milan, 1967).

16. *Ibid.*, Vol. 1, 181, f45r (tavola 86). Next to f45v is Francesco's version of Taccola's hollow tube ram pump. Another pair of grooved ram pumps are on f47v, in this case operated by a centrally pivoted rocking beam. Maltese could not make much of this pump: 'La sua assurdità tecnica è per noi inesplicabile nel quadro della conoscenza pratiche di F di G'. L. Reti, 'Francesco di Giorgio's treatise on engineering and its plagiarists', *Technology and Culture*, 1963, 4i 278–98, reproduces two drawings of grooved ram pumps but without offering to explain how they worked.   .

17. (i) Prager and Scaglia, *op. cit.* (4), 85. (ii) H. Doursther, *Dictionnaire universel des poids et mesures, anciens et modernes* (Brussels, 1840), 72, best of guides, gives sub *bras* two quantities, both cloth measures, for the Sienese braccio: à toile = 23.633 inches, and à laine = 14.867 inches.

18. G. Scaglia, *op. cit.* (11), 1: 109–10. Taccola's language is here unmistakable: 'When the wood (ram) drops the valve shuts and the water is violently discharged' (*eructat* = vomited). I suppose, instead of the scholarly, not to say pedantic, coinage, 'ram pump with cannelure', I might have done better to have called this thing 'the grooved eructor'.

19. J. Needham, *op. cit.* (3), Vol. 5, Part 7, 601, gives bibliographical details of this text (see note 3). It should be noted here that, although the suction lift pump was missing from the Chinese technological repertoire, Needham cites the case of the buckets lifting brine from deep holes in Szechuan as being the closest parallel. These buckets were rather like Taccola's hollow tube pumps but very much longer. They did not, of course, make an airtight seal with the sides of the bore-holes they played in, so they were not really pumps. See Vol. 4, Part 2, 42–3. The self-emptying gaining and losing buckets of Europe were cognate devices.

20. The only contemporary description appears to be that of Adolf Gurlt, *Der Darlington Gesteinbohrer, eine neue Bohrmaschine zum Betriebe von Bergwerken, Tunneln und Steinbrechen . . .* (Bonn, 1875). Dr Hugh Torrens (private communication) writes that John Darlington Junior's (1826–97) drill was first demonstrated in 1873. By 1885 Day & English engineers, of Bath, were advertising their valveless air compressors for driving rock drills. These were being made under a patent granted in 1878 to a certain Edmund Edwards.

21. Leonardo at some time came into possession of a copy of Francesco's work. Like Francesco, he addressed the problem set by the two opposed motions of Taccola's grooved ram pump. About 1488 he devised a brilliant solution to the difficult kinematical problem, that is, how to link elements in circular and straight-line motion so that neither movement constrains the other. For a description of this invention see the author's 'The sector and chain: an historical enquiry', *History of Technology*, 1979, 4; 149–85.

22. Prager and Scaglia, *op. cit.* (4), 39.

23. *Ibid.*, 56.

24. B. Gille, *Histoire générale des techniques*. Vol. 2, *Les origines* (Paris 1958), 154,

also notes this semi-automated mill of Taccola, quoting the latter's very telling phrase, 'Plus valet ingenium quam bubalorum vires' ('ingenuity is worth more than brute force', literally, the strength of oxen).

25. G. Agricola, *op. cit.* (10), trans. H. and L. Hoover (London, 1912; reprinted New York, 1950), 176–7: '(an) iron disc or one of wood . . . is far superior to the 'shoe' ', i.e. the leather cone.

26. T. Ewbank, *A Descriptive and Historical Account of Hydraulic and Other Machines for Raising Water, Ancient and Modern* . . ., New York, 1856, 214.

27. G. Scaglia, *op. cit.* (11), Vol. 1, 110–11.

28. The use of the word 'bucket' for piston in English, as d'Acres used the word in 1659, persisted late into the nineteenth century. It looks like a linguistic skeuomorph, but a comparative historical study of at least Italian, German and French terms for pumps would be a necessary preliminary before any conclusions could be drawn. The problems posed by technical vocabulary are very real. See J. Needham, *op. cit.* (3), Vol. 5, Part 7, 6ff. for a discussion of Chinese gunpowder terminology and the author's 'The vocabulary of technology', *History of Technology*, 1977, 2: 125–55, for a discussion of the term 'Stangenkunst'.

29. G. Simondon, *Du mode d'existence des objets techniques* (Paris, 1989) (1st edn. 1958). See especially ch. 1, 'Genèse de l'objet technique: le processus de concrétisation', and there in particular p. 43, '*L'essence technique* se reconnaît au fait qu'elle reste stable à travers la lignée évolutive mais encore productive de structures et de fonctions par développement interne et saturation progressive'. But first catch your essence, as Mrs Beeton might have said.

Technological Aspirations and the Motivation of Natural Philosophy in Seventeenth-Century England

ALEX KELLER

'Somewhat more space in the monograph has been devoted to the hypotheses about economic and military influences on the range of scientific enquiry than to the hypotheses which link up Puritanism with recruitment and commitment to work in science', observed Merton in his 1970 preface to *Science, Technology and Society in Seventeenth-Century England*.[1] But, he adds, very little has been written since about the former topic, in comparison with the lengthy debate about the Puritan connections of early English science. Indeed, one of the few papers which does give equal space to both themes, A.R. Hall's 'Merton Revisited' (1963) concurs in that point.[2] Has the 'second Merton thesis' been neglected; and if so, why?

One consideration comes to mind. In general, questions of the interaction of technological development and economic demands with the progress of scientific research have not been ignored. Far from it, it would be fair to say that question has been a constant undertone throughout our intellectual history since the days of Bacon. Nearly all late seventeenth-century scientists began by assuming that the pursuit of science would give rise to wonderful inventions that would deliver humanity from the worst of its burdens. It was long a commonplace that contemporary technology had but to apply the discoveries already made by fundamental, curiosity-oriented science. This assumption could always be used to justify expenditure on basic, theoretical investigation, which might seem useless to the lay public. Only quite recently, in the 1960s, has this assumption been challenged. Were technologists really just waiting for crumbs to fall from the scientists' table? Or were they engaged on an independent pursuit of knowledge, also valid, autonomous in practice, and perhaps also motivated by higher aims than turning a quick buck?

At that time it seemed to many that if the resources and the high-flying intellectual manpower that had devised nuclear weapons could only be put to work to solve the problems of peace, the sky would, literally, no longer

76

be the limit. The exploration of space, the medical success of antibiotics, an economic boom powered by cheap fuel all created an air of optimism. Think-tanks and seminars were called into being to explain how science is revolutionizing our life and tell us what further changes in our outlook and way of life are to be expected. But were these the triumphs of science— or of technology? Through the 1960s and 1970s the argument raged— hardly too strong a term, for some quite sharp words were used on occasion.[3]

Quite often this debate drew on contrasting views of the classical Industrial Revolution in England and hardly anything was said about the period about which Merton wrote. Was there perhaps an assumption that if technical development and 'pure science' were effectively still living apart in the eighteenth century that must needs have been even more true of the previous century? Then, even if seventeenth-century scientists were interested in certain subjects for socio-economic reasons, or wished to help the state in time of war, their interest would have been fruitless. Technology would solve some of these problems in the generations that followed: 'science' as constituted in the seventeenth century did not, and could not.

Interest then shifted toward a view of technical knowledge as a 'scientia' in its own right, now converging on scientific knowledge, now diverging, now parallel, but never simply dependent. The technical knowledge could be a simple skill, or it could relate to the theoretical constructs employed in the design of an automated factory. Many, nevertheless, like Hall, have wished to retain a clear distinction between knowledge of the method used to achieve a certain goal and knowledge about *why* that method works, in terms of a fundamental understanding of the materials involved.[4] In that case the latter could be related to the 'scientification' of technological knowledge. This approach is reflected in attempts to seek a philosophy of technology in Germany (e.g. the work of Rapp)[5] and in the United States, notably the Chicago seminar of 1973.[6] The relationship between techno-logical knowledge and scientific theories and discoveries was prominent in these discussions, but little seems to be said about the way technological and economic concerns might have directed the movement of scientific research. Rather, it seems to have been assumed that curiosity-oriented research was directed at the solution of intellectual problems, from which technological benefits might flow; the question was only to what extent technological, artefact-oriented, product-oriented research might claim some independence and intellectual respectability of its own. In the 1970s, German historians and sociologists of science were more exercised by the finalization of science argument raised at the Starnberg Institute for the Study of the Conditions of Life in the Scientific Technical World.[7] They too saw scientific and technological research converging, but agreed that the two modes had had little practical contact in the early stages. The period with which Merton dealt was for the Starnbergers still the explorative, pre-paradigmatic stage of the sciences. Nearly all branches of science were then, they believed, too immature to be able to answer queries about the workings of nature as manipulated by men. So, they too would argue, although Fellows of the Royal Society or the Académie des Sciences might have wished to be useful to the commonwealth, in fact they did not know how to be.

This certainly appears to be the view of Michael Hunter, in his *Science and Society in Restoration England*.[8] Several of a series of lectures delivered at the Clark Library, UCLA in 1978–9, and published as *The Uses of Science in the Age of Newton*,[9] urge this view, which is also reinforced by the conclusions of Kathleen Ochs in her study of the Royal Society's 'histories of trades'.[10] In *The Uses of Science . . .*, Westfall's paper on Robert Hooke[11] remarks that Webster, in his 'massive study of utilitarianism among the Puritan reformers', *The Great Instauration*, discusses five projects. Westfall, however, declares that 'I do not find among them anything to which the word *science*, as I understand it, properly applies'.[12] He himself, therefore, deals with three attempts by Hooke to put his undoubted mathematical and scientific ability to practical use: a lamp with a steady flame, the setting of a sail to drive a ship, and his spring watch. Even in the last case, where Hooke can lay claim to an important invention, Westfall argues that his theory—if that was the theory behind the invention—was quite unsound. Two other papers in this collection deal with topics which Merton chose as examples of the technological interests of the Royal Society. A.R. Hall writes on gunnery and science,[13] and D.W. Waters on nautical astronomy and the problems of navigation.[14] Both agree that science simply could not help the practice of art at that time, or at most indirectly.

In the 1980s, the search for a comprehensive perspective on technical knowledge came to dominate the scene. The seminar on Models of Scientific and Technological Change, held in Pittsburgh in 1981 and subsequently published as *The Nature of Technological Knowledge: Are Models of Scientific Change Relevant?*, provides a lucid view of contemporary debate, but, again, says little on earlier times.[15] In France, too, this was the leitmotif of the latest attempt at a universal history of technology from the Palaeolithic to current events: the *Histoire des techniques*, edited by Bertrand Gille.[16] His own essay on technical knowledge raises some fascinating points, but the concluding section is aptly entitled 'Une situation confuse'. Indeed, to remind us how much this whole discussion depends on categories of language, which have an essential component of the imprecise and the arbitrary, and differ from tongue to tongue, the reader finds that the second part of the *Histoire des techniques* is entitled 'Technique et sciences', but includes chapters on 'Technique et langage' and 'Technique et droit'. Today technology has filled the world with artefacts as Bacon promised; has extended human empire over almost everything that happens on the land surface of our planet; and already pokes eager fingers under the ground, down to the depths of the sea, and beyond our planet's atmosphere. Almost all these techniques have been feasible only because of the steady enlargement and refinement of human knowledge about the internal workings of things animate and inanimate. So it is tempting now to look back and see the first seeding of these conquests in the earliest institutions of modern science. Yet once we look into the question we are forced to accept that the science of the late seventeenth century could provide technologically useful information in very few instances.

Recently Lindqvist has suggested that the transfer of scientific procedures, experimental and quantificatory, from basic sciences to technologies, depended on the development of institutionalization.[17] Only projects 'that were undertaken by institutions rather than individual efforts that were performed within an institutionalised framework' succeeded in 'scientising' industrial processes; that is, in obtaining satisfactory results which could serve to make some, at least, scientific. Only institutions could exercise sufficient control for adequate periods of time over manpower and materials to 'reproduce laboratory conditions in the field'. But, adds Lindqvist, we should look for institutions in which the authority to deploy these resources was based on 'scientific and technical competence'. At first glance the early Royal Society and similar bodies, like the Académie des Sciences and the Accademia del Cimento, might seem to fit the bill. In practice, however, proposed 'histories of trades' were desultory collections, the work of individual enthusiasts and amateurs, lacking authority and certainly lacking competence in the fields they investigated. Perhaps that is one reason why these 'histories' often had to wait a long time for publication and had little impact on science, and even less on the processes of production.[18]

So, if we can suggest why Merton's 'research related to socio-economic needs' failed to answer those needs, can we suspect that is also why there has been little interest in finding out whether that research really was related to those needs? As Merton has commented, in all the literature on how science affects and has affected society, much less has been said on the way society affects the development of science. The endeavour to follow this less popular path led him to ask whether the blockages perceived in particular flourishing and expanding branches of the economy may have encouraged them to research this question rather than that. He agreed that scientific curiosity might take over, that scientists may have come to enjoy the intellectual chase so much they lost sight of the utilitarian interest of their quarry. In any case utilitarian rewards might with hindsight seem unattainable. But, at least, utilitarian motives might have first attracted them.

This could be regarded as a milder version of the classic exposition of the Marxist view set forth in Hessen's paper on 'The Social and Economic Roots of Newton's *Principia*'.[19] At one point Hessen claims that 'in order to develop its industry the bourgeoisie needed science which would investigate the qualities of material bodies and the forms of manifestation of the forces of nature'.[20] He gives a shopping list of the bourgeoisie's requirements: e.g. the 'construction of canals and locks demands a knowledge of the laws of hydrostatics, the laws governing the efflux of liquids', or, a couple of pages further on, 'Ventilation equipment demands the study of draughts, i.e. it is a matter of aerostatics', while pumps, we are assured, combine the two, and need 'considerable investigation in the realm of aero- and hydrostatics'. All very well, but, as has been remarked above, could aerostatics and hydrostatics in the age of Newton have told the miner anything about ventilation that he did not already know, or the builder of locks more than empirical experience had already taught him about the behaviour of water when discharging through pipe, sluice or leat? Outstanding technical achievements of the late seventeenth century, like the

Canal du Midi or Dutch wind-powered manufactures, seem rather the cul-
mination of skills that had been accruing for centuries. Among the few
genuinely radical inventions of the period, most were concerned with the
expansion of knowledge rather than production. If, later, procedures and
approaches first employed on scientific apparatus were carried over to
industrial uses, that does not affect their original purpose. The high technology of the seventeenth century was embodied in telescope, microscope,
barometer, thermometer, wind-gauge, pendulum clock, air pump. Even
the micrometer and the universal joint, which were to have so many industrial applications, were first devised to make telescopes more efficient.

There is, of course, a more thoroughgoing version of the Marxist
view, according to which, as Hessen explains, 'the method of production
of material existence conditions . . . the intellectual process of the life of
society'. Does that not mean that the bourgeoisie determines the answers
as well as setting the questions? Merton's argument is more modest; he
simply insists that the contemporary economic situation sets the agenda.
Once seventeenth-century men had decided that the new natural
philosophy was a fit matter in which to engage their lives and efforts, they
had in theory an almost unlimited number of topics to explore. Those which
they actually investigated were thrust upon their attention by an expanding
economy, or by the military demands of expanding empire, not to mention
civil wars. Merton, therefore, looks in depth at mining, transport and
military technologies. Despite their differences in interpretation, he does
adopt much of Hessen's analysis.[21] However, he also accepts the criticisms
which Clark had made of Hessen's case[22] and, adds Merton, 'with this
revision, I am in substantial agreement'. Since Clark had maintained that
besides economic life and war, science was also influenced by the arts,
religion and, 'above all, the disinterested search for truth', that does leave
the 'second Merton hypothesis' rather up in the air.

Unlike Hessen, Merton did carry out a detailed personal research into
available data. His attempt to quantify those of the early Royal Society's
researches that were 'related to socio-economic needs' as opposed to pure
science, however crude, was a pioneer effort, and admirable for that very
reason. Before (and all too often since), so much of this debate has been
fogged with terms so woolly they could neither be proved nor disproved,
or it relied on anecdote or reports dredged up selectively from muddy
waters. Since Merton's work, others have tried to quantify prosopographical
data about the early Royal Society and the topics that exercised it. Unfortunately, problems of definition lay so many pitfalls in the investigator's
path that safe and secure conclusions are hard to achieve. In one of the first
enquiries to concentrate on the topics rather than the people, going through
the same data as Merton, B. Staff decided that 'the Fellows appear to show
more willingness to discuss pure sciences and natural curiosities, than to
study technology with the necessary persistence for success'.[23] The
magisterial work of Hunter does confirm this, as indeed does Ochs.[24]

If we go through Merton's own chapters, doubts do arise. The coal industry is given pride of place in the chapter on mining, although almost all the
techniques mentioned were, at that stage, interesting because of their use

in metal mining.[25] Mine drainage is, Merton believes, the most important field of research inspired by the development of coal. However important it might have been to drain mines so that they could be worked, we should have expected that mine captains and capitalists would also be concerned about more efficient techniques of extracting coal or ore. The ancient tradition of fire-setting to split the rock survived unchanged. It was a purely empirical technique that demanded much skill and nerve but little science. In the seventeenth century this method was supplemented by the use of gunpowder to blast the rock. That must have been an even riskier operation, whose practitioners built up over the years a sense of the right place to set their charge, the right quantity of powder, and the safe distance from which to fire it. Presumably this process was transferred from the well-established use of gunpowder to undermine enemy fortifications. Once the rock came away, miners broke off and broke up lumps of ore or coal with picks, whose geometry and material had been much improved over the ages, but were really not so much more efficient than the antlers of neolithic flint miners. So perhaps drainage was the first aspect of mining to be mechanized because there was already a repertoire of water-raising apparatus, inherited for the most part from Greco-Roman antiquity, if not older still. These devices had been invented to get water out of wells or from streams, most often for irrigation. They were then transferred to the mines. Suction pumps had been a later development, perhaps as recent as the fifteenth century. These raised a problem indeed, since they could only lift water to a limited height. The plungers could be adapted to lift water directly, as well as draw it up under atmospheric pressure, so that the pump was not restricted to the height to which water might be forced by that pressure. Even so, suction pumps, like force pumps, suffered from problems of leakage, bursting of pipes, and strain on various parts, increasingly so as miners ventured to greater depths. German engineers had solved that problem sufficiently for their own purposes by linking pumps through a series of connecting rods, the famous *Stangenkunst* which could also provide a means to link the pump to relatively distant sources of water power.[26] Curiously, this device did not catch on in the England of the early Royal Society. Still, analysis of ideas that were proposed suggests that the problem was seen as a choice of mechanism, permuting diverse forms of 'simple machine' to raise a given body a given distance with a given power. The solution actually came from envisaging a source of power that was not given—the steam engine. Of the scientists, only Papin thought of that. It is true that steam was introduced in order to exploit the pressure of the atmosphere, the 'ocean of elementary air', as Torricelli called it, and that was certainly a scientific discovery.

In the chapter on transport, it is presumed that the development of oceanic navigation, and so intercontinental trade, depended on knowing a method of finding the longitude precisely.[27] Did shipmen need to know this? Certainly, it would have been useful. Yet, by the time states took an interest in the matter and even offered money rewards for a solution, was it to promote commerce, or for their navies which had to seek out, or avoid, enemies? Charles II did grumble that the Royal Society was not spending enough time working for his sailors. Civilian shipping might always take advantage of an unexpected landfall, and indeed had started up more than one new line of trade that way. By the time a chronometer worthy of compensation had been developed, virtually all of the coastlines of the world's great land masses had been explored and regularly visited by Europeans, as had many of the world's larger islands. Only Arctic and Antarctic shores eluded them—but not for the difficulty of finding them. Slowly, regular oceanic journeys removed all the cartographic ghosts which faulty estimates of longitude had produced in early sixteenth-century maps, and removed them without much assistance from scientists. In practice the concept of geographical coordinates, as well as the basic technology of using this concept to locate oneself on the earth's surface, were both ancient, and both were gifts of astronomy. Seamen did indeed adopt, adapt and simplify the instruments of the astronomers, while the astronomers sometimes tried to think of something new. Again, we are dealing with attempts to find new applications for traditional devices. The astronomers had not studied astronomy in order to find a better way of finding longitude at sea. Using celestial observations for the purpose of location was an essential of their stock in trade since earliest times. Now they felt they could make themselves useful. Of course astrology too was, they were convinced, a useful art which often inspired them. It does not seem that merchants or mariners complained to the astronomers; indeed, they were rather inclined to object that the astronomers were quite impractical, and sometimes poked fun at their interference. Significantly, the answer came not from new, improved principles of celestial observation, as the astronomers had supposed, but from the work of clock-makers who found out empirically how to build a timepiece protected against all the variables of motion at sea—temperature, humidity, pressure—which the scientists could not then quantify. Besides, there was the practical problem—a problem for craftsmen rather than scientists—of protecting instruments from the corrosive effect of minute granules of sea salt, carried in fine spray or dripped from wet sleeves on to moving components.

If socio-economic motives had been primary, or even political ones, would not the scientists have turned to researches into how stronger, swifter ships could be built, or their crews kept healthy? There was the occasional experiment in this line, such as William Petty's 'double-hulled' ship, a kind of catamaran. Such ventures were rare and seldom successful. Hessen declares that 'In order to improve the floating qualities of a vessel it is necessary to know the laws governing the movements of bodies in liquids'; and a ship's stability 'is one of the basic tasks of the mechanics of material points'.[28] When Newton in the *Principia* says of Proposition XXXIV, Book II, 'this . . . I conceive may be of use in the building of ships', are we to suppose that he had been led into the study of the resistance of fluid media to the motion of spheres and cylinders for that purpose? Hall indignantly comments that at that time 'no master-shipwright employed mathematical theory or would have admitted the competence of a mathematical physicist to instruct him'.[29] Without Newton's help or advice, however, they built ships, aided only by scale models and drawings, which could cross oceans and carry their crews to far-off destinations through storm and calm.

If we turn to Merton's final chapter on socio-economic technological science, that on the military, similar points can be made.[30] These pages are dominated by research into ballistics. Now ballistics was indeed the 'New Science' set before the world by Tartaglia in 1537, perhaps the first truly novel branch of physics since antiquity.[31] In the next generation Benedetti busied himself with this topic, and so did Galileo.[32] But Tartaglia presents himself as urging his science on unwelcoming and incurious gunners. His ballistics arose not from their requests but from Aristotle's theories of motion and the problems which those theories had caused. Proposals had been made centuries before to avoid the inconsistencies of Aristotle's exposition. Among these inconsistencies, projectile motion was the most troubling. But neither Tartaglia nor Galileo needed cannon to make them think about projectiles. Human beings have been shooting arrows since long before *they* were born—indeed long before Aristotle.

Why did ancient Greek armies not encourage the study of ballistics to enable them to shoot better? After all, the Hellenistic kingdoms did finance research into the construction of more powerful weapons. So why did nobody bother about a ballistic theory to explain how missiles fired from a ballista landed where they did? Rather than believe that Galileo went into dynamics to find out why cannonballs hit or missed their targets, all the evidence suggests that he wanted to explain how bodies on or close to the earth's surface share the motion of a rotating earth; this enquiry led him to his 'law of free fall', and to the remarkable gestalt switch we call the concept of inertia. From these abstract concepts he developed a theory of projectile motion that could be consistent with the rest of physics.

Afterwards, it evidently occurred to him that he could now improve on the range tables that had been published by Tartaglia's followers. In practice, as Hall has shown, such tables, including those by Galileo's successors such as Torricelli, were almost useless.[33] When neither scientists nor gunners had any means of assessing the muzzle velocity of a cannonball once fired, when nobody knew why that particular combination of chemical substances thrusts the ball forth so violently when ignited, when smooth-bored cannon were far from smooth—how could a science of ballistics tell the artilleryman more than he knew by experience? So guns continued to be laid by technicians who had had to learn the feel of their tools, helped by the simplest of mathematical rules of judging distance and height. They usually hit their targets—eventually.

So it might be suggested that the three topics in which Merton finds strongest evidence of social and economic 'push' towards particular lines of enquiry were all taken from longstanding branches of scholarly knowledge. The learned, who had taken these sciences up for a variety of psychological motives of their own, did hope to impress society with the utility of their studies, and wished to persuade the rest of society that 'virtuosos' were not abstruse scholars chasing private will o' the wisps, but had a very practical contribution to make.

Even in those days, moreover, teaching and instruments cost money. If the Royal Society and the Académie des Sciences had tiny budgets by modern standards, they did have to raise those sums by asserting their value to the mercantilist state. As Laudan remarks, 'If the wish is father to the deed, then the hopes of generations of supporters of science for a technological payoff from (and hence justification for) their scientific research might well in itself have been adequate to generate the myth' that technology is, or should be, the application of science, and that any science will one day lead to technological spin-off.[34] There could of course be technical spin-off from a scientific idea or discovery found in the search for answers to theoretical problems. When Galileo found the satellites of Jupiter, he was not looking for a solution to the longitude problem. But having observed them, he did realize that the occultations of four satellites as they moved around their mother planet would make a better check on terrestrial position than the motions of our Moon. He began to draw up tables of these satellite positions, and his efforts won him a golden collar from the government of the Netherlands (which he had to turn down). But he did not get very far. Many years later, during the period studied by Merton, Romer endeavoured to draw up more exact, and so more useful, tables. In the course of observation he stumbled on anomalous effects, which led him to discover and measure with some degree of accuracy the speed of light. But he had hardly been looking for that, either.

Have we perhaps been misled by allowing developments since the middle of the nineteenth century to cast a backwards shadow? Hessen observes that Newton's mechanics had to be restricted to exchanges of kinetic energy, or, as Hessen puts it, 'one form of movement (mechanical)'.[35] Now many other forms of energy operate in our environment and in ourselves. But only after industry had made use of thermal energy through the steam engine, and then chemical and electrical forms of energy, could we come to understand that all were interchangeable, and energy conserved through diverse transformations. Hessen's claim does sound reasonable. Until technologies had evolved so as to give humans control over these exchanges beyond the kinetic, scientists were unlikely to be able to make fundamental sense of them. So the new technologies would be the *sine qua non*, if not the driving force. Their processes gave the scientists fresh puzzles to solve. By the mid-nineteenth century at all events, industrialists might turn to science for increased efficiency or improved materials. If scientists' own motives were undoubtedly still mixed, intellectual curiosity could well be reconciled by then with expectations of economic benefits.

Yet what of the exothermic chemical reaction that is the explosion of gunpowder? Within limits, surely that was a technique under human control in the seventeenth century. However, the state of chemistry did not allow scientists to calculate how the explosion was transformed into the mechanical motion of the cannonball. An acceptable explanation came after the comprehension of what went on in heat engines, not before. Although steam power entered the world long after guns, its efficiencies and inefficiencies could be more easily perceived and measured.

Perhaps in the history of steam, too, the spin-off factor may be allowed a role. The discovery of atmospheric pressure was the work of physicists who were not looking for new sources of power, nor even for barometric instruments. They wanted to produce an artificial vacuum in order to shake the assumption that nature abhors a vacuum, which had been the linchpin of Aristotle's argument against atomism. The failure of suction pumps to raise water above a clearly defined height gave Galileo a clue. But was it his objective to design a pump that would raise water higher? Surely he and Torricelli, Pascal, Von Guericke and all the others were drawn to this question by its relevance to the atomic theory of matter, to which they were all committed. Von Guericke saw the force which the atmosphere exerted upon an evacuated vessel. In the age of Newton, scientists like Huygens, Hautefeuille and Papin experimented with 'vacuum' or atmospheric engines, employing gunpowder. That led nowhere. Only Papin seems to have pursued the question with persistence, turning from gunpowder to steam. Unfortunately we know so little about the education and intellectual contacts of Savery and Newcomen, who constructed the first working steam engines, that conclusions must be tentative. Evidently ideas about vacua, and about the pressure of air and 'vapours', must have been quite widespread through the population. Even Newcomen, who was probably more of the unlearned tradesman than Savery, must nevertheless have become aware of the proposition that it was due to atmospheric pressure that water could rise so far and no further in the suction pump, and that atmospheric pressure might be exploited to lift it.[36]

All the same, technical men might well be interested in scientific theories. The German contemporary of Boyle and Hooke, Rudolph Glauber, was a pioneer industrial chemist in the manufacture of drugs. He was also an innovator in chemical theory. Did he choose those theoretical fields because they might cast light on his technical problems? His theories do look quite loosely connected to his practice, although he had a profound intellectual curiosity and tried to explain what he had discovered sometimes in the search for new and potent drugs. In his case the theoretical constructs are dead as doornails—but Glauber's Salt is still on sale.

Could there be other connections between the technologies of a progressing economy and the pursuit of science? Price, in his contribution to the Pittsburgh symposium, seeks a link in 'instrumentalities'.[37] In his view shifts in the paradigms of science frequently result from 'a change in the technology of science which may be rather trivial and is almost always an intruder from some vastly different current in the history of technology'. Such instrumentalities 'may be a substance, an effect, a phenomenon, a methodology, a technique', which does widen the term instrumentality to include almost anything but pure thought. His prime examples are Galileo and the telescope, Volta and current electricity. Perhaps he does stress too much the accidental, serendipitous element in their discoveries. Galileo probably did have some notion that with his telescope he would find something that could serve his intellectual programme, even if he could not foresee what. Once invented, the telescope became a tool of intellectual enquiry; technology, hardware, in the service of theory. 'Whilst the merchant companies turned to him for his telescope,' declared Hessen, 'academic philosophy turned a blind eye.'[38] In fact, after some huffing and puffing the learned world accepted the telescope with enthusiasm. But when did telescopes become standard issue in the navies of Europe, let alone the merchant ships? Not for another century at least. Price, however, would have it that the telescope was the unintended by-product of the manufacture of lenses: 'Once the lens grinding lathe had been developed for the mass production of eyeglasses, deepgrinding a lens blank became easy.') So, highly concave lenses could be produced in some quantity. Perhaps the legend that a spectacle-maker's children at play in his shop discovered the magnifying power of a pair of lenses, if not true, is at least *ben trovato* and encapsulates a deeper truth.

Or does technology, fruit of human need, just supply models for theoretical constructs? That would seem to agree with what had been said of the history of energy. Many years ago Grossmann argued that science was shaped in the image of technology and of society through language, through its terms and metaphors, and visualized models.[39] The machine itself begot the physics of the age of Newton. To be sure, the new scientists of the time spoke on every occasion of their 'mechanical philosophy'. Among the more naïve the imagery of clock or machine was taken literally, with references to springs, levers, wheels within wheels, and so on. Grossmann himself identifies four main types of machine which inspired this philosophy; firearms, waterworks, lifting mechanisms, clocks. He would have the story go back to Leonardo da Vinci; but all the mechanisms and devices he mentions were much older than Leonardo. Clocks, in practice, were the preferred model, the most familiar to the professional classes and the most intricate and complex mechanical artefact then known in Europe. Once wound the clock operates automatically, one simple, moving component pushing or turning the next. It breaks up the uninterrupted flow of time, as we subjectively feel it, into precisely designated and equal units which are translated into the simplest movements of one body upon another. The idea of a clockwork universe has indeed become so obvious and so well known that it has served as the title of exhibitions.

If we claim that without machines, and in particular without clocks, there is no mechanical philosophy, are we saying more than Hall when he jokes about *Principia* Prop. XXXIV of Book II: 'Newton could not have written these words if he had been unaware of things called ships'? As it developed, the mechanical philosophy left clocks and other machines behind. Even Newton himself, on occasion, had to defend his physics and his cosmology against the charge that he was not mechanical enough. The metaphor became more and more metaphorical.

In the end, Hall in 'Merton Revisited', and Merton revisiting himself in 1970 would probably agree that we have to reject 'the mock choice between a vulgar Marxism and an equally vulgar purism'.[40] Hall perhaps enjoys trailing a coat at times. In *The Revolution in Science 1500–1750*, the best general guide to early modern science,[41] he certainly allows a good deal to 'some technical influences', although it is doubtless significant that he has left this chapter virtually unaltered from his original formulation of thirty years before. Proponents of more uncompromising views are obliged to gloss over contrary evidence. It is now agreed that socio-economic problems of the forces of production did not dictate the choice of research topic; still less did they dictate the results of research.

If it is so difficult to maintain a definite connection between the three technologies which Merton held as tokens of prime importance and the actual problems chosen by leading minds of the early Royal Society, and if the procedures and the information acquired by those investigations were not very serviceable, what then are we left with? I should like to propose standing Merton on his head. His principal thesis argues that recruitment, commitment and legitimation for the pursuit of science were linked to a Puritan outlook, while technology provided the research topics. I should rather ask: did not the original recruitment and commitment to science come from technological enthusiasm rather than from Puritan inspiration? If we consider the most productive brains of the early Royal Society, those who really made the scientific discoveries, several of them set out on that road with excited hopes that a world of helpful knowledge waited only to be revealed. A heady cocktail of Bacon, Comenius and Hartlib, drunk by youthful stomachs empty of direct acquaintance with the industrial crafts, turned their brains. Later they had to sober up . . . but by then an enthusiasm for knowledge as a good thing in itself had taken over.

Of these bright lights of the Society several were very young when first illuminated. Boyle was eighteen or nineteen at most when he met Hartlib in London, and soon after got involved with Worsley and his 'invisible college'.[42] Wren and Hooke were both students at Oxford when Wilkins drew them into his circle of experimental philosophers.[43] Wren's huge catalogue of some fifty-three 'new theories, inventions, experiments and mechanic improvements exhibited . . . at the first assemblies at Wadham College in Oxford' must date to his early twenties at latest—some probably before he was twenty.[44] They include a few ideas related to Merton's three topics, and even the odd industrial invention, such as a device to 'weave many ribbons at once with only turning a wheel'. But Wren carried out few of these schemes in later life. Hooke, too, was full of ideas as a student, and doubtless his work on his spring watch and the air pump encouraged him to propose a utilitarian approach to the Royal Society in its gestation. Petty was twenty-five when he published his *Advice* . . . for Hartlib on the establishment of a scientific college, roughly on the lines of Bacon's Salomon's House.[45] If Wilkins inspired Hooke and Wren, he was himself a little older when he met Hartlib, and his eyes were opened. His path in life had always lain in the Church. From the time of their contact Wilkins turned to the popularization of Galileo's astronomy. Then in 1648, in his *Mathematical Magick*, he fairly bubbles with ingenious ideas—submarines, flying ships, sailing carriages, perpetual lights. Most, however, were quite impractical and many, indeed, he took from his reading of Cardano or Mersenne.[46] He, too, was keenly convinced of the prospects for innovation and improvement through mathematics but had little real experience of what was involved.

Amid our talk of industry, navigation, mining and war, we should not forget that a number of the most active founders of the Royal Society thought mainly of medical techniques when they envisaged the benefits to mankind of scientific research. The Halls have reminded us, as has Robert Frank, not to forget the medical interest in the establishment of the Society.[47] Not only were medical men active in the world of this infant science, but some of those whom we would not now regard as doctors had had a medical training. Viscount Brouncker had a medical degree from Oxford, even if he is remembered chiefly as a mathematician, and for his involvement with the Navy Board in the early years of Charles II, which led him to experiment on the recoil of guns.[48] Petty had studied medicine abroad and became professor of anatomy at Oxford, while John Evelyn, the driving force behind the 'histories of trades' project, had studied medicine at Padua. Abraham Cowley, whose *Proposition for the Advancement of Philosophy* appeared just as the nascent society was taking shape and helped to publicize it, had qualified in medicine not long before.[49] Not surprisingly then, pharmaceutical questions were of greater concern to members than industrial ones; presumably also of greater concern than military matters, although there the Fellows' wish to demonstrate their utility to the state may have weighed more with them.

Ochs remarks that when deciding which trades to investigate, 'these scholarly gentlemen did not focus on economically important industries, but rather on those related to biology and chemistry'.[50] Perhaps it is somewhat anachronistic to speak of biology at that period; better say medicine and agriculture and horticulture. The latter was certainly Evelyn's forte, and his intellectual progeny are to be sought among the improving landlords of the next century rather than the mechanical inventors; the 'Turnip' Townshends more than the James Watts.

In Petty's imaginary college, skilled craftsmen were to be encouraged to come together to a 'gymnasium mechanicum' where they would live rent-free, and through their collaboration and exchange of ideas 'all trades will miraculously prosper and new Inventions would be more frequent than new fashions of clothes and household-stuffe'. However, a much more prestigious and more important element in his proposal is the 'Nosecomium Academicum', a medical school to teach anatomy, surgery and chemistry, with its botanic garden, zoo, and museum of natural history as well as technology.[51] This enthusiastic belief that medical and technical innovations would follow from a better understanding of nature survived for some years through the early 1660s, approximately up to the time of the Great Fire of 1666, and no doubt was a major factor in the actual foundation of the Royal Society. Perhaps their self-confidence was undermined by the shock of seeing one of the most powerful and flourishing capitals of Europe helpless before two successive catastrophes, Plague and Fire. However that may be, Boyle, Hooke, Wilkins and Wren made little effort to realize the wonderful inventions of which they had dreamed in the 1650s. Wren turned to architecture, and Wilkins to linguistics, with his search for a rational and universal language.

Although he returned to it much later in life, Petty's double-hulled boat also belongs to the years before the Fire. Thereafter he turned his attention exclusively to political and economic issues. After 1665–6 Hooke's remarkable technical versatility was devoted in the main to instruments for scientific observation. I would not wish to claim that all the prominent scientists of mid-seventeenth-century England were similarly inspired. John Wallis, whose autobiography is a valuable document for their history, makes no suggestion that he had been thinking on those lines when he joined Wilkins and his friends in Civil War London. Although 'navigation' and 'mechanicks' are mentioned among the sciences which he says that company discussed, the actual topics he lists do not include any of a technological nature. Interestingly, his only practical work (apart from decoding documents) was his study of the sounds made by the human voice, which he tried to apply in teaching the deaf to speak.[52]

This interpretation of an early enthusiasm and later loss of interest, even while the Society was still quite active in the pursuit of natural knowledge, is borne out by analysis of the articles in the Society's Museum, described in the catalogue by Nehemiah Grew published in 1681.[53] By that time the majority of the collection comprised specimens of natural history. Of the 'artificial' items, most were what we would now regard as ethnographic curios of distant peoples. This was, to be sure, not just out of interest in other cultures, in lands with which Europeans were only now becoming familiar. In the tradition of Bacon and Hartlib, these new scholars did not yet assume, as later generations were to do, that British technologies were so superior to those of the rest of the world that they could have nothing to learn from them. Such remote nations might indeed possess skills and materials which the English could usefully copy. The artefact section is divided into three: instruments of natural philosophy, instruments of 'mathematics and mechanics', and what would now be termed fine arts (preserved in this particular collection on account of their ingenuity rather than their beauty). Much of the section headed 'Mechanick' was non-European, apart from a box-bellows, the gift of Sir Robert Moray. Evelyn had donated three items for the 'husbandry' section: a Spanish combine plough–sower–harrow; a saffron kiln; and a cider-press. Navigation is represented by a model of Petty's boat and by a couple of instruments designed by the young Wren. From Moray came a novel hive which young Wren had 'contrived' in 1652. There were a few military devices, including Prince Rupert's gunpowder assayer, Brouncker's gun fixed in a triangle, for his recoil experiments, and a marvellous musket which fired seven shots in succession, with automatic reload of bullet and refill of powder. This had been invented by Dudley Palmer, who died in 1666. The scientific instruments formed a large category and included Newton's reflecting telescope, one of the last artefacts to enter the collection. Many of the others were Hooke's work (such as the lamp analysed by Westfall). Most items of this type had come from Wilkins. Only three English items can be associated with anybody other than the persons named. Although hardly any are dated, it is clear that very few had been added during the 1670s.

It is true that the *Philosophical Transactions* still accepted 'technological' articles for publication. The role of validation which the Society performed through its journal could serve inventors also. So Papin showed the Society his steam digester in 1679, and Savery his steam engine twenty years later.

Just recently, Bryden and Simms have examined in detail how the Royal Society came to endorse the products of a new technique to grind optical glass truer, and in larger quantities, which the London spectacle maker John Marshall submitted to them in 1693.[54] In these cases, however, the initiative seems to have come from the inventor seeking the Society's approval, to assist in his publicity; the Society did not sponsor them or try to steer its members' own research in that direction.

The Fellows themselves, certainly the most able of them, had come to concentrate on natural philosophy, on experiments of light, in Bacon's phrase, with little expectation that their work would lead to fruitful experiments. Long ago, old Aristotle had insisted that the wisest and most rational of human endeavours is the desire to know how the world works. When we reason our way to an understanding we are most human, and yet most god-like. Petty—a hardheaded and practical man if ever there was one—claimed to one of his friends, 'I had rather live on herb pottage all the days of my life (as I did with advantage at the time of my sickness) than not to study truth and those symmetries whereby the world stands and which are the causes why "res nolunt male administrari".'[55] Even those who came to the study of natural philosophy inspired by an emotional commitment to enquiries that would lead to use, through the control of nature, gradually came to realize that nature would not easily be controlled. Since they were still convinced that nature might be understood, however, they were satisfied that the search for those truths and symmetries was motive enough.

Notes and References

1. R.K. Merton, *Science, Technology and Society in Seventeenth-Century England* (New York; Harper & Row, 1970), xii.

2. A.R. Hall, 'Merton Revisited', *History of Science*, 1963, 2: 1–16.

3. This debate is surveyed in A.G. Keller, 'Has Science Created Technology?', *Minerva*, 1984, 22: 160–82.

4. A.R. Hall, 'On Knowing, and Knowing How To . . .', *History of Technology*, 1978, 3: 91–103.

5. F. Rapp, *Analytical Philosophy of Technology*, (D. Reidel, Dordrecht, 1980).

6. Published as G. Bugliarello and D.B. Doner, eds., *The History and Philosophy of Technology* (University of Illinois Press, Urbana, 1979).

7. Published as *Die gesellschaftliche Orientierung des wissenschaftlichen Fortschritts*, English trans. enlarged, W. Schaefer, ed., *Finalisation in Science* (D. Reidel, Dordrecht, 1983).

8. M. Hunter, *Science and Society in Restoration England* (Cambridge University Press, 1981).

9. J.G. Burke, ed., *The Uses of Science in the Age of Newton* (University of California Press, 1983).

10. K.H. Ochs, 'The Royal Society of London's History of Trades Programme: an Early Episode in Applied Science', *Notes and Records of the Royal Society of London*, 1985, 39: 129–58.

11. R.S. Westfall, 'Robert Hooke, Mechanical Technology and Scientific Investigation', in Burke (9).

12. *Ibid.*, 89.

13. A.R. Hall, 'Gunnery, Science and the Royal Society', in Burke (9), 11-141.

14. D.W. Waters, 'Nautical Astronomy and the Problem of Longitude', in Burke (9), 143-69.

15. R. Laudan, ed., *The Nature of Technological Knowledge: Are Models of Scientific Change Relevant?* (D. Reidel, Dordrecht, 1984).

16. B. Gille, ed., *Histoire des techniques* (Gallimard, Paris, 1978); English trans., *The History of Techniques* (Gordon & Breach, London, 1986).

17. S. Lindqvist, 'Labs in the Woods: The Quantification in the Late Enlightenment', in T. Frangsmyr, J.L. Heilbron and R.E. Rider, eds., *The Quantifying Spirit in the Eighteenth Century* (University of California Press, Berkeley, 1990) 291-314.

18. Ochs (10) is the most recent study of this question. I trust I have fairly interpreted her conclusions. She does make it clear that the programme helped to set in motion 'the transfer of manufacturing knowledge from the craftsman to the engineer, scientist and corporation'—largely to the disadvantage of the former—which in the long term led to modern industrial production.

19. B. Hessen, 'The Social and Economic Roots of Newton's *Principia*', in *Science at the Crossroads*, papers presented to the International Congress of the History of Science and Technology, 1931 (new edn, Frank Cass, London, 1971).

20. *Ibid.*, 170.

21. As he acknowledges; Merton, *op. cit.* (1), 142, n. 24.

22. G.N. Clark, *Science and Social Welfare in the Age of Newton* (Oxford University Press, 1937; new edn, Oxford University Press, 1970), particularly the third chapter.

23. B. Staff, 'The Place of Technology in the Early Royal Society 1660-1990', (M.Sc. thesis, University of Leicester, 1960). These are the concluding words of the summary, but cf. also pp. 76-7.

24. Hunter (8), and also his *Establishing the New Science* (Woodbridge, 1989), and Ochs (10).

25. Merton, *op. cit.* (1), 137-59.

26. G. Hollister-Short, 'Leads and Lags in Late Seventeenth-Century English Technology', *History of Technology*, 1976, 1: 159-83.

27. Merton, *op. cit.* (1), 160-83.

28. Hessen, *op. cit.* 158-9.

29. Hall, *op. cit.* (2), 8.

30. Merton, *op. cit.* (1), 184-98.

31. N. Tartaglia, *La Nuova Scientia* (Venice, 1537).

32. S. Drake and I. Drabkin, *Mechanics in Sixteenth Century Italy* (Wisconsin I.P., 1969), 224-8.

33. A.R. Hall, *Ballistics in the Seventeenth Century* (Cambridge University Press, 1952).

34. R. Laudan (15), 9.

35. Hessen, *op. cit.* (19), 193-203.

36. L.T.C. Rolt, *Thomas Newcomen: The Prehistory of the Steam Engine* (Hartingdon, 1977).

37. D.J. de S. Price, Notes towards a Philosophy of the Science/Technology Interaction, in Laudan, *op. cit.* (15), 105-14.

38. H. Grossmann, English trans., 'The Social Foundations of Mechanistic Philosophy and Manufacture, *Science in Context*, 1987, 1: 137-80.

39. Merton, *op. cit.* (1), xiii.

40. A.R. Hall, *The Revolution in Science 1500-1750* (Longman, London, 1983), revised from *The Scientific Revolution* (Longman, London, 1954).

41. *Ibid.*

42. R.E. Maddison, *Life of the Honourable Robert Boyle* (London, 1969), particularly 67–73.

43. B. Shapiro, *John Wilkins* (Berkeley, 1969), ch. V, 118–47.

44. S. Wren, *Parentalia, or Memoirs of the Family Of Wrens* (London, 1750, facsimile edn. Farnborough, 1965), 198–9.

45. W. Petty, *Advice of W.P. to Mr Hartlib for the Advancement of some particular Parts of Learning* (London, 1648). These ideas are discussed by W. Houghton, 'The History of Trades: Its Relation to Seventeenth-Century Thought', *Journal of the History of Ideas*, 1941, 2: 33–60.

46. J. Wilkins, *Mathematical Magick; or The Wonders that may be performed by Mechanical Geometry* (London, 1648).

47. A.R. and M.B. Hall, 'The Intellectual Origins of the Royal Society— London and Oxford,' *Notes and Records of the Royal Society of London*, 1968, 23: 157–68. B. Shapiro and R.G. Frank, *English Scientific Virtuosi in the Sixteenth and Seventeenth Centuries* (University of California, Los Angeles, 1979).

48. J. Dubbey, 'Brouncker, William' in *Dictionary of Scientific Biography*, 1970, 2: 506–7; J.F. Scott and H. Hartley, 'William Viscount Brouncker', *Notes and Records of the Royal Society of London*, 1960, 15: 47–56: cf. also A.R. Hall, 'Gunnery, Science and the Royal Society' in Burke, *op. cit.* (9).

49. A. Cowley, *A Proposition for the Advancement of Natural Philosophy* (London, 1661; facsimile edn, Merton, 1969).

50. Ochs, *op. cit.* (10), 136.

51. Petty, *op. cit.* (45), 7–8.

52. C.J. Scriba, ed., 'The Autobiography of John Wallis F.R.S', *Notes and Records of the Royal Society of London*, 1970, 35: 17–546.

53. N. Grew, *Museum Regalis Societatis; or a Catalogue and Description of the Rarities Belonging to the Royal Society* (London, 1681).

54. D.J. Bryden and D.L. Simms, 'Spectacles Improved and to Perfection and Approved by the Royal Society'. I am most grateful to the authors for their kindness in showing me a preprint of this paper.

55. E. Strauss, *Sir William Petty, Portrait of a Genius* (London, 1954), 166.

Count Theodore Batthyány's Paddle-Wheel Ship

WALTER ENDREI

The late Roman writer, the Anonymus author of *De rebus bellicis*, wrote his treatise about AD 375. He proposed, among other things, that the empire should supplement its shortage of manpower in its struggles against its enemies beyond the imperial *limes* with machines. One of his proposals was to use animals, or more precisely oxen, to propel ships by means of paddle wheels and thus dispense with large crews of oarsmen. The most recent editor of the treatise considers that the pictures accompanying the MS of the work preserved in the Bodleian Library, Oxford, the Codex Oxoniensis Canonicianus Lat. Misc. 378, reproduce with substantial accuracy the designs of the Anonymus.[1] These miniatures were probably painted about 1430 and were the work of German illustrators working in the tradition of the Upper Rheinish school. The manuscript and its miniatures were prepared for Pietro Donato, Bishop of Padua, and were in his hands by January 1436. The Anonymus's animal-powered ship is shown in Figure 1.

Figure 1 *De rebus bellicis* miniature of *c.* 1430: ship with three pairs of paddle wheels driven by oxen.

93

By then the idea had already gained fairly wide currency and had appeared, for example, in the work of Konrad Kyeser of 1405, and by 1472 was in print for the first time in Valturio's work, *De re militari*. Even Leonardo did not disdain to discuss this subject: of course the authors of the fifteenth century always postulated human muscular force as the driving agent.[2] Even the paddle-wheels of the first large vessel of 200 tons, shown to Charles V in the harbour of Barcelona, were turned by twenty-five to forty men in 1543.[3] Later projects, like those of Ramelli, of Prince Rupert of the Rhine whose vessel overtook the Royal Barge of Charles II on the Thames, or of Papin, had no noteworthy consequences either.

It is interesting to note, however, that the number of patent applications increased considerably towards the end of the eighteenth century in almost all countries. The reason for this may be sought in the increased demand for transport capacity due to the dynamic increase of population and the Industrial Revolution, and in the well-known difficulties of towing—boat mills, towing paths. One of the attempts at a solution should be mentioned here: P. Miller in 1797 constructed a double boat with the paddle-wheel arranged between the two hulls (boat mill), which won a race against sailing boats in the Firth of Forth. In the next year, however, the inventor replaced human muscular force with a steam engine built by William Symington: this is the oldest boat engine in existence, in the Science Museum, London.[4]

Count Theodore Batthyány (1729–1812) was the younger son of the last Royal Palatine of Hungarian lineage. His brother was Archbishop of Esztergom. He studied in Vienna, mechanics among other subjects, and soon developed a lively interest in technical problems. This was manifested not only by his opening up and buying mines and manufactures of cloth, china, majolica and needles, but also by his establishing a large technical and natural science library.[5] He also studied river control problems in great detail. In 1774 the Court ordered him to develop a plan for regulating the river Kulpa, a task which he undertook gladly, since he had large estates in Croatia for growing corn (see map). For the same reason he set up a small boatyard in Ozály on the Kulpa where a number of sailing boats were built.[6]

E. Süsz[7] in 1777 was the first to give news of the arrival of one of the largest vessels of Batthyány and a contemporary, J.M. Schweighofer, wrote in 1783 that 'Count Batthyány has appeared twice already with a large galley before the gates of Vienna, attracting the eyes of the people and the attention of trade experts . . .'[8] This ship, copied from a Rhineland type by the constructor Happe, but with a length of 52 metres and a capacity of 8000 centners (4.5 tons), was the first 'Bucentaurus' which made the voyage between Vienna and Croatia two or three times every year. He sold two ships of similar size to the Willeshovenian Trade Company, which later also bought the Bucentaurus to establish a direct Vienna–Cherson connection.[9] It was at this point that Batthyány began to be interested in an invention that would make possible 'upstream and downstream navigation in rivers'. Why so late? It is difficult to find the motive for this. In the opinion of Siklóssy and Biró the announcements of such inventions,

Map: Hungary, the Danube and its Croatian tributaries.

appearing with increasing frequency, had stimulated his ambition. In the *Gemeinnützigen Blättern* (Papers of Public Interest), a Miller of Pressburg was mentioned who had apparently invented a boat capable of 'gliding easily' upstream. In the same year (1785) Márton Holló, a university professor from Nagyszombat, started building a similar vessel: the gazette *Magyar Hirmondó* wrote about his paddle-wheel boat in 1787.[10] Similar news items reached him, probably from Western European sources, for had he not been able to send his brother a drawing of the Montgolfier brothers' balloon less than two months after their first ascent in it?

However, in this context I must turn back to the inventor's library. This contains several works on navigation, some of them valuable old tomes, such as the *Tractatus de Mercatura* by Benvenuto Straccha (Venetiis, 1575), in which the chapters 'De navibus' and 'De navigatione'[11] furnish detailed information. There are also publications from Batthyány's time which are quite modern works. One of these seems especially instructive. It bears the title, *Notice of a New Invention for Going Upstream with a Loaded Boat* (Vienna, 1786). This booklet presents in detail the status quo of Hungarian river navigation and also mentions the Bucentaurus of Batthyány; however, it stresses the invention of the author without giving any details, except to say that he was looking for a partner to realize it.[12]

Obviously, therefore, this anonymous author could have been a probable companion when Batthyány in the early 1790s applied for an 'exclusive privilege'.[13] This application does not contain a detailed description of the boat; it states that 'Count Bathiani . . . has expended extremely great intellectual efforts, care and costs' upon it, and repeatedly mentions a companion without actually naming him.

However, the construction of the boat and also the granting of the privilege were slow to come about. Batthyány transferred his shipyard from Ozály to Pressburg (Bratislava), but at the coronation of the Emperor Francis in June of 1792 he appeared in Vienna only on a magnificent raft. The patent was issued only in 1793.[14]

The 'Votum' for issuing the patent contains some remarkable statements by an official which are cited below:

> I cannot decide whether the petitioner has much to fear from the copying of his ship. It seems to me to belong to the same class as the ship in Leopolds-Stadt which is driven upstream by fire.

This indicates that by this time early patent applications for steamships had been submitted in Austria, as they had in the USA. The unknown author of the text finally remarks with a certain malice:

> Since the petitioner undertakes not to hinder present-day navigation, of which I have no doubt, perhaps he might be permitted to enjoy the pleasure of expending even more money for an Imperial privilege.[15]

This application, submitted during the reign of Leopold II (the Votum bears the date 2 September 1791) was apparently refused as Reichspatent, and the two privileges issued to him were drawn up for Transylvania (6 February 1793) and for Hungary (17 June 1793) and were valid for twenty years. Each of them contained three conditions:[16]

1 Within three years the ship should be investigated by experts with respect to establishing its applicability.

2 To prevent any litigation regarding the patent, an exact plan or a model should be submitted.

3 The patent should not serve as a pretext to prevent the development of actual navigation or shipbuilding in the country.

The ship was completed—probably in the summer of 1793—and presented to the public, as attested by an article in the *Magyar Kurir*.[17] The eyewitness describes the ship as oval in shape, completely covered, but with an open gallery surrounding it. The access door to the interior could be pushed up and down. The ship had a flat bottom, but no paddle-wheel or rudder were visible. Later Batthyány complained that the shipwrights had made a mistake in building the ship and that certain parts of it had to be reconstructed.[18]

Whatever was the actual fact, we are only informed about the partial destruction of the ships—for he had built several—'by the push of the ice'. This occurrence has been represented in several water-colours, together with the interior space arrangement of the ship and the triumphant presentation of the flotilla.[19]

This latter event occurred in September 1797 'in the august presence of

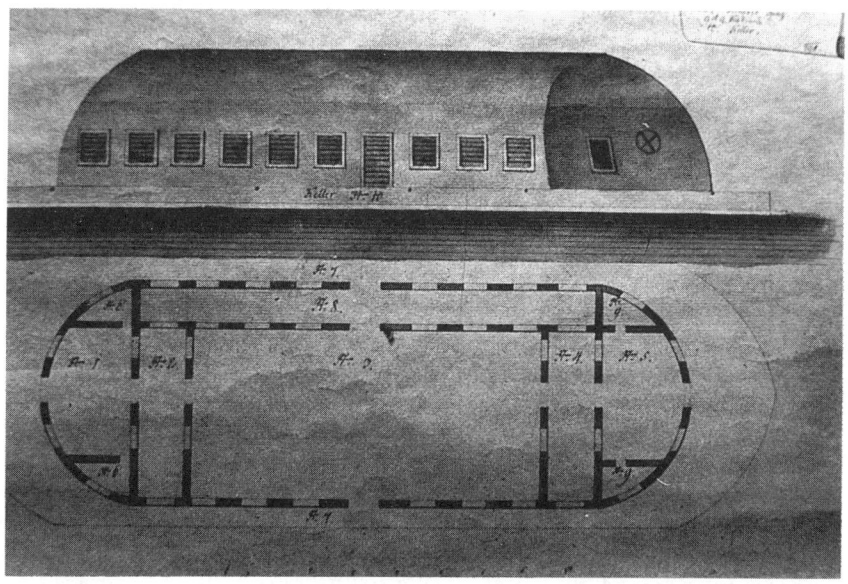

Figure 2 Outer appearance and outline of one of the smaller ships.

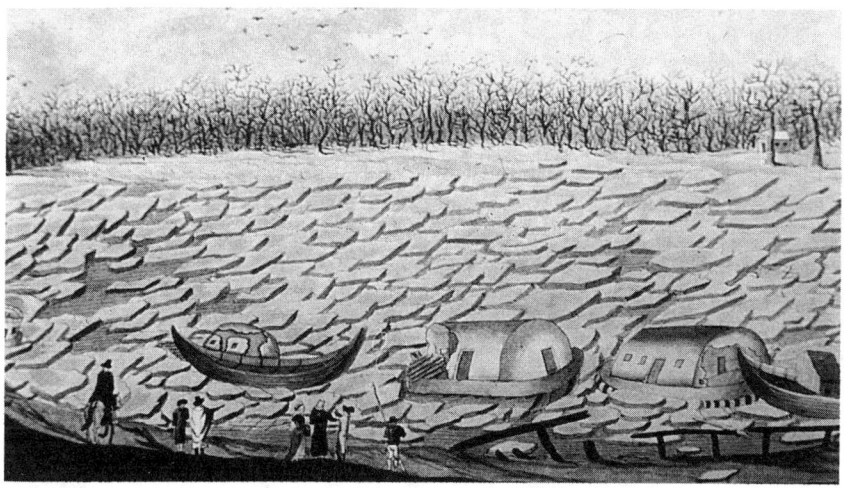

Figure 3 The wreckage of the flotilla near Pozsony (1796).

Figure 4 The presentation of the flotilla on the Danube near Vienna: in the background the spire of St Stephen's Cathedral is visible.

their Majesties the Emperor and Empress, and of many grandees of the Reich and of a countless number of admiring people', in Pressburg. The main vessel had a weight of 700 centners (39 tons), and it towed two smaller ships and moved upstream without any difficulty.[20]

In the following year, 1798, the new Bucentaurus was converted and in August four of these novel ships came to Pest. The *Pressburger Zeitung* on this occasion furnishes details of the driving of the ship. The reporter mentions 'the completely newly invented ship for parade and diversion . . . the like of which has never been seen on any river in the whole world'. It might be 'called also a freight ship' and, although no bigger 'than 4 Kellhamer (Kehlheimer), it may be pulled upstream with fewer horses'.[21]

From this last sentence one might suspect that the ships were towed upstream. However, the plans and illustrations show no devices at all for attaching ropes. The text continues enigmatically, stating that 'the vessel travelled upstream solely through the influence of the river itself, with the aid of a moving machine invented specially by His Excellency'. It has similarly 'travelled upstream publicly in Vienna, under the Tabor Bridge and along the Augarten into the Brigitta-Aue'.

Mention is made also of the fact that the count had left the ships in Vienna, since he had plans for building further vessels on his estate, Siklós, on the Dráva. However, one may speculate on the type of boat 'driven solely by the influence of the river'. This might point to a ship-towing device (remorquage) but there are absolutely no proofs and preconditions for its existence.[22]

The description of the ship stresses the salon, decorated with chandeliers and mirrors, but also mentions a hold with a length of 17½ fathoms

Figure 5 Reconstruction of a ferry-boat driven by two horses walking on a turntable. *National Geographic Magazine*, 1989, Vol. 176.

(33 metres) which, with a width of more than 12 metres, could have readily accommodated one or more winches and paddle-wheels. The height of this cellar space must also have been considerable, at least 2 metres. Especially noteworthy seems the aperture in the side of the ship through which fragmentary details of the ship's driving mechanism may be glimpsed.

We know that such vessels, driven by winches, functioned without any problems in the nineteenth century in spite of the steamships beginning to evolve, mainly as ferries on river and sea crossings in the USA. On the frontier between the states of New York and Vermont in the period 1814–58 at least seven such ferries may be traced; an 1820 engraving of Philadelphia shows two such vessels in operation, and not far from the Niagara Falls such a ferry connected the Canadian shore with the US shore. A photograph of such a horse-driven winch has been preserved. The last ferry, driven by a blind horse, was running in the state of Tennessee until 1929.[23]

At the beginning of the Napoleonic wars there seemed to be good prospects for Batthyány's ship. He joined the Royal Hungarian Navigation Society, which in 1800 obtained a privilege for maintaining ship traffic on the Kulpa 'with vessels of 300 and even 600 centners to promote the supply of our army in Italy'.[24] The peace of Lunéville (1801) abruptly brought this project to a halt.

Figure 6 Panorama of Philadelphia (1820) showing a horse-driven ferry-boat. Mariners Museum, Newport News, Virginia.

The Bucentaurus is mentioned once more. Apparently it had gone from Vienna to Pressburg in June 1802 and a series of festive events were held on board.[25] A rare pamphlet describes its arrival in an 'allegorical-historical story' of nine stanzas. From this one may conclude that the Imperial couple were present at the landing.[26]

By this time the inventive ambitions of the ageing Batthyány were exhausted. In 1810, at the age of eighty, he wrote to his brother from Paris, 'I am living quietly and alone, far from all foolishness and I have said goodbye to all extraordinary ideas.'[27]

Notes and References

1. (i) E.A. Thompson, *A Roman Reformer and Inventor, Being a New Text of the Treatise 'De rebus bellicis . . .'* (Oxford, 1952), 16–17.
(ii) F.M. Feldhaus, *Die Technik der Vorzeit, der geschichtlichen Zeit und der Naturvölker* (Munich, 1965), 26 'Anonymus de rebus bellicis'.
2. (i) *Ibid.*, 936–9, 'Schiff mit Schaufelrad'.
F.M. Feldhaus, *Leonardo der Techniker und Erfinder* (Jena, 1913), 123–7.
3. Feldhaus, *op. cit.* (1, ii), 938.
4. *Ibid.*, 923ff. and 939ff.
5. W. Endrei, 'Batthyány Tódor Müszaki konyvtara' (The technological library of Theodore Batthyány), *Könyvszemle*, 1991.
6. J. Biró, 'Batthyány Tódor hajóépitö és hajózási kisérletei' (Theodore

Batthyány's shipbuilding and shipping endeavours), *Közlekedési Muzeum Évkönyve*, I, 1971, 239–64.

7. E. Süsz, 'Die Bedeutung der Donau', *Streffleurs Österreichische Militärische Zeitschrift*, Vienna, 1885, Bd 26, No. 1, 8.

8. J.M. Schweighofer, *Versuch über den gegenwärtigen Zustand der österreichischen Seehandlung*, Vienna, 1783, 44.

9. Biró, *op. cit.* (6), 46–8.

10. L. Siklóssy, 'Batthyány Tódor gróf ár ellen haladó hajója' (Count Theodore Batthyány's ship moving upstream), *Búvár*, 1940, 97; Biró, *op. cit.*, (6). 248ff.

11. Library of the Hungarian Academy of Sciences, MTA Kt, Régi Gazd., 0.91.

12. MTA Kt. 560.469.

13. Haus-, Hof- und Staatsarchiv, Vienna, Privilegien, 'Batthiani', 251.

14. Biró, *op. cit.* (6), 250–2.

15. *Loc. cit.* (B), 253. The 'Votum' author's mention of 'the ship in Leopoldstadt which is driven upstream by fire' refers to a project of a German constructor (name unknown) who addressed a memoir to Maria Theresa towards the end of 1778 proposing the cannibalization of the derelict steam engines at Schemnitz as a source of equipment for his idea of steam-driven ships to navigate up the Danube. For further details see E. Kurzel-Runtscheiner, 'Die ersten Versuch einer Dampfschiffahrt auf der Donau . . .', *Beiträge zur Geschichte der Technik und Industrie* (Berlin, 1928), Vol. 18, 69–72.

16. State Archive Budapest, P 1320, S/312 and Helytartótanács Dep. Commerciale, 1793, fons 152, pos 1.

17. *Magyar Kurir*, 8 October 1793.

18. Biró, *op. cit.* (6), 255.

19. In the possession of the Museum of Transport and of the author responsible.

20. *Pressburger Zeitung*, 1797, 935; *Magyar Hirmondó*, 1797, 422.

21. *Pressburger Zeitung*, 1798, 777–8.

22. F.M. Feldhaus, *op. cit.* (1, ii), Spalte 'Schiffstauerei'.

23. D.G. Shomette, 'Heyday of the Horse Ferry', *National Geographic*, October 1989, 176: 549–56. R. Selkirk, 'Animal Powered Paddle-Ships', *Archaeology Today*, 19 January 1988, 10–11.

24. Országos Széchenyi Könyvtár, Hungarian National Library Manuscripts, Fol. Germ., 1575/1.

25. *Pressburger Zeitung* (1802), 570.

26. *Der siegende Bucintor* . . . (Pressburg, 1802).

27. Biró, *op. cit.* (6), 263.

The Kongsberg Silver Mines and the Norwegian Mining Museum

BJÖRN IVAR BERG

ABSTRACT

The silver mines at Kongsberg in Norway (1623–1958) were of major importance in the eighteenth century. These technologically most advanced mines in Norway applied German techniques, with an extensive use of water-wheels for drainage, hoisting and processing. A water-supply system with dams and aqueducts is largely preserved in addition to many other features such as wheel-house walls, inscriptions, mine shafts, rock waste heaps, footpaths, etc., the whole forming a fascinating early industrial landscape. The underground workings some with installations are also quite well preserved. Significant are the abundant curved forms resulting from firesetting, a technique which was used extensively until 1890 for breaking the predominantly hard rock. Research in the history of technology and the communication of this and other historical aspects to the public is the field of activity of the Norwegian Mining Museum. The execution of these tasks is facilitated by the rich archives and collections of maps, drawings and objects, as well as by the presence of mine workings, workshops and the industrial landscape.

INTRODUCTION

Pre-industrial mining was characterized by early industrial traits such as machines, non-manual energy, professional engineering, division of labour and work discipline. This is not a matter of interest only for scholars. Many material remnants are preserved: tools, machines and other equipment, workshops and buildings, and underground mines. This cultural heritage offers an instructive arena for the communication of the history of technology to a wide public. The distinctiveness of the arena and of the actual techniques attracts attention. Although some are complex, many of the techniques with their concrete and elementary mechanical features are intelligible to the non-technical public.

Three aspects facilitate this communication:

1 The presence of surviving materials ranging widely both in form (tools, machines, buildings, mines and time, illustrating technological change at different stages of development to full industrialization;

2 Knowledge of these historical processes, obtained through research based on both the remnants themselves and on written sources;

3 The possibility of combining and ability to combine the surviving materials and knowledge into on-the-spot communication. It is of great value to have accessible sites presenting technology in its authentic environment.

The former silver mining town of Kongsberg, Norway, has rich endowments to support these aspects. The task of preservation, research and communication based on the history of the silver mines is the field of activity of the Norwegian Mining Museum.

THE KONGSBERG SILVER MINES

Mining for copper in the area is recorded back to 1490, and is the subject of the earliest written source on metal mining in Norway of uncontested reliability (there is a possible reference to a small silver mine in Oslo in the late twelfth century). Silver mining was started seriously in 1623. In the eighteenth century, Kongsberg became a major European silver producer, with the peak of production being reached in 1768 with 8258 kg, and a working staff, in 1770, of 4000 employees. The average annual production in the eighteenth century was 4435 kg of silver.[1] Following a decline, most of the mines were closed in 1805. Shortly after the reopening in 1816, rich deposits were discovered, and the mines operated profitably until the late nineteenth century, although on a much smaller scale than previously. The silver mines were finally closed in 1958. Throughout their 335 years of operation they had been a state enterprise except for 43 years in the seventeenth century. Before 1814 Norway, and hence the silver mines, were under the Danish crown. The total production was about 1,350,000 kg of silver, and the estimated total input of employed work[2] about 300,000 man-years (250,000 of them before 1800). This was the largest mining enterprise in Norway ever measured in terms of human work input.

AN OUTLINE OF THE TECHNOLOGICAL HISTORY OF THE SILVER MINES[3]

A geological and structural note
Mining took place in a large number of both small and large mines, on deposits consisting of calcite veins, mostly a few centimetres wide, with very variable silver content. The veins had a steep dip and were silver-bearing in rather short zones, giving the mines the form of vertical shafts enlarged to stopes of lengths varying from a few to around 100 metres (Figure 1). In addition there were numerous 'search' (exploratory) headings and some long drainage adits.

Excavation of rock and ore
The main technical fields of mining were excavation of rock and ore, transportation, and draining the mines of water. Excavation mainly

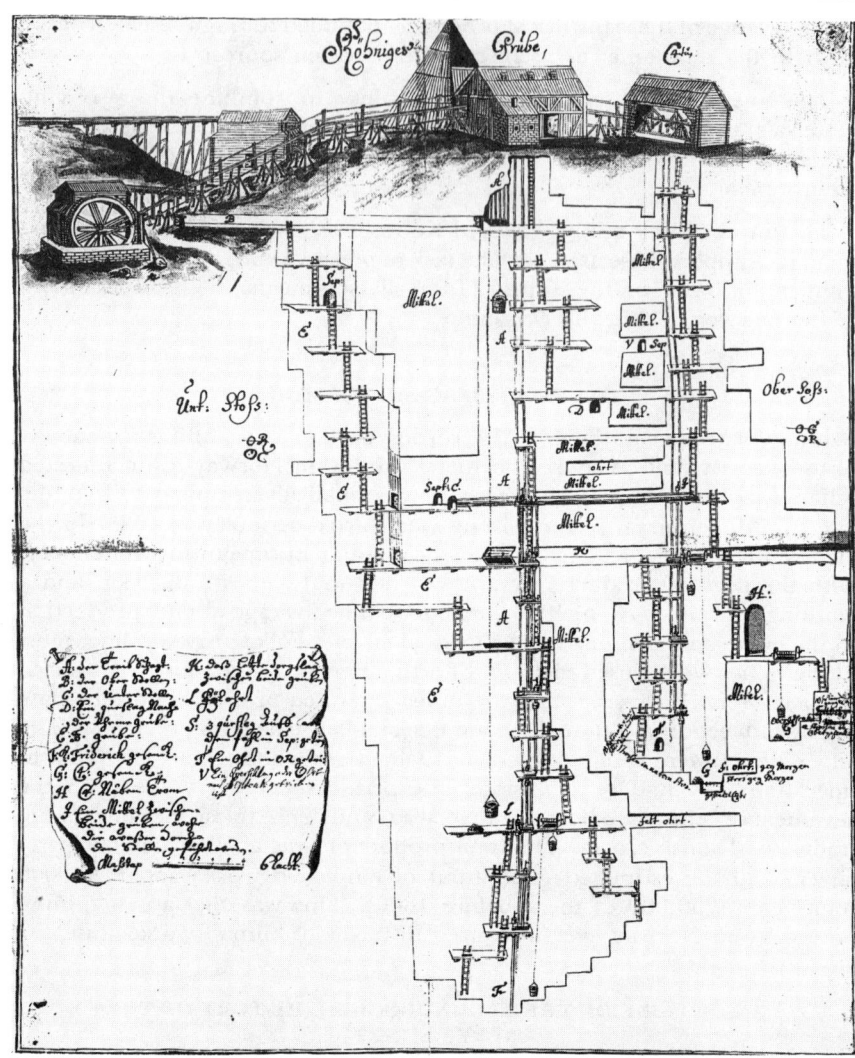

Figure 1 Vertical profile of the King's mine, *c.* 1715. The mine, opened in 1623, has reached a depth of 115 m. It has three shafts from the surface, with massive rock ('Mittel') in between. A horse whim in the conical building is hauling ore buckets in the central hoisting shaft. Rock and ore are brought by three hand winches from the bottom of the shafts. Drainage is by two rod-engines placed on the surface with field-rod transmissions which operate pumps up to the lower adit level ('Sept: C.'). The upper adit ('B'), already completed in 1624, can be documented in the archive records. The adit was made with firesetting and is in part preserved. (Map at the Norwegian Mining Museum)

consisted in breaking side rock, owing to the nature of the deposit. This involved removing hard, silicious rocks not well suited to the ancient techniques of hammering with chisels and iron wedges, although these techniques were used during the first century of mining at Kongsberg. Another ancient technique, breaking rock with heat through firesetting, was more efficient in dealing with hard rock. Firesetting was also applied from the start of mining, and within forty years became the dominant technique. An alternative was blasting with gunpowder, the most important innovation in seventeenth-century European mining, diffusing throughout Europe after its first successful demonstration in Slovakia in 1627.[4] Although it was suggested in Kongsberg in 1643, tried from 1659, and a more serious attempt made to employ it from 1681, considerable problems with this new technique hampered its large-scale introduction until the first third of the eighteenth century. Only after crucible steel replaced forged iron as drill material from 1860, and especially after the introduction of dynamite from 1874, was firesetting finally abandoned in 1890. Excavation was not mechanized till 1908, with the introduction of compressed-air-driven drilling machines.

Water drainage

As far as mine water drainage was concerned, however, mechanization had already taken place two decades after the start of mining. The actual technique was the rod-engine (German *Stangenkunst*), which by then was established as the standard European drainage machine, developed between the mid-sixteenth and early seventeenth centuries. A rod-engine consisted of an overshot water-wheel as motor, power transmission through a chain of connected wooden rods, often traversing the surface terrain (field rods, *Feldgestänge*) as well as in the shafts, linked in the shaft to a series of piston pumps (Figures 1 and 2).[5]

Some of the water-wheels, like the first that was set in operation in 1644, were placed in the mineshaft itself, but most of them were built on the surface with double field-rod transmission, the first of these in 1646. This machine system had been fully developed in central Europe shortly before its introduction at Kongsberg, including switchgear to change the field rod's direction in order to overcome terrain fluctuations. The largest system from the 1640s had a water-wheel of diameter 10 m with a main double field-rod of about 380 m in length with 58 simple 'swings' (between the rods), and 3 switchswings (for vertical changes of direction), all resting on 92 logs and 46 ramps, transmitting power to a central pumping shaft serving four mines.[6]

A large number of rod-engines were built, and this machine, although sometimes in the variant run by a horse whim, remained the standard drainage machine at Kongsberg until 1871, when water-column engines, working force pumps, were introduced. The last rod-engine stopped work in 1900.

Transport of rock and ore

Because of the physical structure of the mines, transport of rock and ore was mainly a matter of shaft hoisting until railway traction, from 1842,

Figure 2 The Norwegian Mining Museum, pump from the Bratte prospect, *c.*1750. This suction-lift pump is of the classical rod-engine type, although in this instance it was run by a horse whim. The pump cylinder is made of cast iron. To the left is the piston with its rod and the iron hook which connected it to the shaft rods. On the far left is a pipe, and in front a part of the deepest pipe, protected from damage through blasting by iron rings. Most of the rods and pumps in this mine had been preserved in water until the reopening of the mine in 1936. (BVM-F 320026)

made horizontal haulage through adits an economic proposition. Shaft hoisting could be done manually longer than water drainage, because the weights to be transported were less. Handwinches sufficed for a long time, especially in the early era of face working with hammers and chisels. With the increased use of firesetting after 1660, the masses of rock waste increased as well as the hoisting distances, with a progressive deepening of the shafts, and, as a consequence, horse whims were introduced in 1670 (Figure 1).

Figure 3 Kongsberg with the mining hill after the establishment of the water-supply system. Close to the mines there are horse whims and wheel-houses with field rods. Dams and aqueducts can be seen near the mines and under the peak of Jonsknuten. From *Museum Regium* (catalogue of the Royal *Kunstkammer* in Copenhagen), 1696.

The first water-wheel for hoisting was introduced as late as 1727. This was a reversible wheel (*Kehrrad*), but this first truly mechanized hoisting became important for large-scale mining only later in that century. Reversible wheels were also used until 1900.

Energy supply
The crucial point in seventeenth- and especially eighteenth-century mining at Kongsberg was the energy supply. About fifty overshot water-wheels with diameters between 9 and 13 metres were in operation for drainage and hoisting purposes, with a number of water-wheels in ore-processing plants in addition. The mines were situated on a hill with lower terrain dropping away on all sides and lacked streams or brooks to run the wheels (Figure 3). The miners were compelled to rely on rainwater falling within the area, and as much of the precipitation as possible had to be collected, stored and properly distributed through an extensive system of dams and aqueducts (Figures 4 and 5). The severe climate, with long dry periods during the cold, icy winters and hot summers, demanded the establishment of storage capacity to supply water throughout the year.

The water-supply system, originating from the 1640s, was designed on a grand scale from 1686, under a new German management recruited from the silver mining district of the Harz, where a similar system had been constructed in the Zellerfeld area during the two preceding decades.[7] In the late eighteenth century the Kongsberg system reached its full extent with

Figure 4 Henrik's dam with the peak of Jonsknuten. The dams are built of two rock walls with a sealing layer of peat between. The outlet regulator is within the small house. This dam, possibly named after the *Berghauptmann* and leading designer of the water system, Heinrich Schlanbusch (1686–1705), was built in 1692. By 1838 it had been modified three times. (BVM-F 319727)

about sixty dams and networks of about 50 km of aqueducts (minor side-systems included). Many of the dams were small and the storage capacity was about 1.5 million m^3, far less than for the three other major European systems, but Kongsberg had nearly as many dams as the largest system in the Upper Harz.[8]

One of several distinctive features of the Kongsberg system was that the water-wheels were mainly placed overground. Three interconnected branches of aqueducts from the dams on the mountain ran their own series of water-wheels in descending lines, where the first wheels at the highest elevations ran the longest field rods for pumping as they had to be placed far up on the hillside above the shaft openings. This was also the case with the reversible wheels for hoisting. It was, in fact, only the introduction of the combination of reversible wheels and transmission through field rods that made the extensive mechanization of hoisting possible. Although the reversible wheel had been known since the fifteenth century,[9] and field rods since the 1560s,[10] this combination was first successfully introduced in 1697–8 at the Great Copper Mine of Falun, Sweden, by the famous engineer Christopher Polhem.[11] The innovation came to Kongsberg in 1734, with a new German technical manager, via the route Sweden–Harz–Norway—a detour characteristic of the significant

Figure 5 Aqueduct at Louisehaug, representing the 10 km of leats which are still maintained and conducting water. This most sophisticated type of aqueduct carries the water completely inside a wooden construction. Outside there is a layer of peat and usually, since the aqueducts run along hillsides, a supporting rock wall. (BVM-F 308137)

German influence on Norwegian mining in general and Kongsberg in particular.

THE NORWEGIAN MINING MUSEUM AND ITS COLLECTIONS

The museum was founded as the Silver Mines Museum in 1938 on the initiative of one of the mine overseers, and was located in the old refinery (Figure 6). It was a company museum as long as the mines were in operation, and remained under the supervision of the Ministry of Industry

Figure 6 The buildings of the Norwegian Mining Museum and the miners' church, which dates from 1761. The dominant building is the silver refinery from 1844, with the machine house to the left and the ore storage house to the right. Far right: a reconstructed crushing plant. Above the refinery: the small balance house and behind it the laboratory. (BVM-F 320008)

until 1965, when it was transferred to the Ministry of Church and Education (later Ministry of Culture) and designated as a national museum of mining. In 1989, the museum was transformed into a foundation and achieved more autonomy resulting in an expansion of resources and activities. One of the new facilities open to the public is a 5 km walk through the *Underbergstollen* adit, an eighteenth-century drainage gallery made by firesetting (Figure 7).

In 1992 the museum had 85,000 visitors in all. It has a staff of ten regular employees, and in addition a number of seasonal guides, employment-assisted staff, volunteer retired people, etc., are employed. The museum also houses the Royal Mint Museum and the Kongsberg Ski Museum, presenting the great era of Kongsberg's ski jumpers. A transfer of the Kongsberg Armory Factory Museum to the Mining Museum is planned.

When the Silver Mines Museum was founded, collections of old tools and equipment had started in some of the mines when they were reopened for renewed mining. These mines, which had been closed around 1800, provided many objects now forming the core of the rich collection of mining equipment that the museum possesses today (Figure 2). As some mines were still in operation when the museum was established, more recent mining techniques, those of the early twentieth century, are also abundantly represented in the collection.

Figure 7 The *Underbergstollen* adit, an eighteenth-century drainage gallery made by firesetting, which is evident from the arched profiles. A 5 km long guided walk through the adit is a popular new attraction of the Mining Museum. (BVM-F 319209)

Collections, exhibitions and guiding of tourists in the mines have a tradition predating the foundation of the museum. The silver mines participated in many world exhibitions starting from the first one in London in 1851, as well as in other, later, international and national exhibitions. Some of the showcases are preserved. The major objects in these early exhibitions and collections were specimens of native silver, crystallized in wires and other spectacular forms. Today, its unsurpassed collection of native silver probably constitutes the greatest attraction of the museum. The abundance of such specimens has made Kongsberg a famous site among mineral collectors. Such specimens were also highly desired as souvenirs or rarities by visitors to the mines, and the accounts registering specimen sales to persons from the 1620s onwards give evidence of this early tourism.

ARCHIVAL MATERIALS

The accounts of the silver mines are preserved for each year from the opening in 1623 until the closure in 1958. They are important sources for the history of technology, giving detailed documentation of dates, descriptions, costs, effects and the people involved. However, the utilization of these sources is time-consuming, as a single year in the mid-eighteenth century may run to about 15,000 pages of accounts.

There is also much other material of great value in the archives: regular records such as correspondence, routine plans and mine descriptions, and also more special sources like voyage reports, discussions on technical problems, etc. There is a large collection of mine and surface maps and construction drawings. This must be one of the most valuable company archives from the pre-industrial age. It has benefited from the official character of the company, the managers also being the head state officials of mining in Norway, and the archive thus being also the archive of the Mining Collegium (*Overbergamtet*). Among the interesting records connected with this institution are the court books, as each mining region had its own jurisdiction. They give many personal and other details. An official archive closely linked to that of the silver mines is the archive of the Financial Collegium (*Rentekammeret*), later the Ministry of Finance, the supervisors of the *Overbergamtet* and of the direction of the silver mines.

Since 1962, most of the archive has been kept in the National Archives in Oslo (*Riksarkivet*), but as a new state archive is now being built in Kongsberg we hope to regain the silver mines archive in due course. Together with other archives from mining it will help to make Kongsberg the research and documentation centre for the history of mining in Norway. We are also working to have the silver mines library returned, also removed thirty years ago, which consists of a rich collection of international mining and scientific literature of the seventeenth, eighteenth and nineteenth centuries. This literature is important to have at hand in the historical research of this extremely detailed and, to many, rather alien field.

German is the dominant language of the older parts of the archive, as well as in the library collections, demonstrating the significant German influence on the Kongsberg silver mines. Correspondence and reports are mainly in German until around 1720, and some series of accounts run in German until 1756 when the last German-born *Oberberghauptmann* died (the last German-born director died in 1869).

THE PRESERVED MONUMENTS AND EQUIPMENT

On the surface, mining has combined with nature to form a landscape of significant features. Neatly placed in the open landscape of the transition zone between bare mountain and forest are the lichen- and moss-clothed rock walls of a number of dams with connected aqueducts that collect and store the precipitation from higher altitudes (Figures 4 and 5). From the dams, three branches of the old aqueducts, amounting to runs of about 10 km, still lead water to the mining area, where the high rock walls of

Figure 8 The oldest of the ninety also preserved wheel-house walls, at the Herzog Ulrich mine, built by Bastian Koth in 1649 for a water-wheel of 10 m diameter with a 126 m long field rod to the mine. The wall is very low compared to those common in later times, and this is probably evidence of an adjustment of German technology to the cold Norwegian winter climate. (BVM-F 308125)

nearly ninety wheel-houses represent the former consumption points of the energy system (Figures 8 and 9). Because of the cold climate, the water-wheels had to be completely enclosed with rock walls reaching to crank level and timber buildings upon them, the latter long since removed or decayed.

Unlike many such places, the mines stand open to view from above. The solid rock has protected the mines from collapse, and their shafts have only in exceptional cases been closed. Some mines have preserved their seventeenth-century narrowness; others are large, formed by the extensive blasting during the eighteenth century of low-grade ore in side-veins, hoisted with water power. The characteristic curved profiles of firesetting are abundant (Figure 10). Rock waste heaps tell of the relative size of the different mines and constitute major features of the landscape. The legendary, but only occasionally fulfilled, optimism of the miner is testified by a large number of prospections. There are perhaps a thousand mine shafts and prospections. The landscape tells the history of mining, mainly before 1800, its development from the early time of rich ores in the pre-blasting period, to the poorer times of low-grade ores mined with new technology. In the landscape there are also a number of inscriptions hewn into the rock in memory of silver finds, great technical plants, etc. The old network of small roads and footpaths is extensively used for walking tours by the people of Kongsberg.

The vertical shaft systems can be visited only by using climbing

Figure 9 The biggest wheel-house wall is this one from 1837, for the reversible water-wheel at Gottes Hülfe in der Noth mine. The wheel was 16 m in diameter and its axle rested on this rock wall, upon which a large wooden building was placed. (BVM-F 308136)

techniques. But at certain levels, many of them are also accessible through adits. The deepest and most important mine, the King's mine, also has a lot of equipment preserved, which can be seen by visitors taking the museum train 2.3 km through the Christian VII adit into the King's mine, 342 m below the surface.

The most interesting machine in this mine is the man-engine or *Fahrkunst*, an elevator consisting of two parallel wooden rods with platforms permitting an effortless descent and ascent (Figure 11). This technique was invented, according to one story, by a pump watcher in the Harz mines who during his shifts had to climb the ladders alongside the *Stangenkunst* rods, so he found himself an easier way by hammering nails for footsteps into the rods.[12] After observing this arrangement, an official in 1833, Georg Dörell, built the first proper *Fahrkunst*, and the innovation was adopted in a number of mines in Europe and America.

The Kongsberg *Fahrkunst* was the only one built in Norway and is one of very few preserved. Unusually, it was run by a water-column engine, which has also, again unusually, been preserved *in situ*, although in the late 1960s the engine was connected to an electric motor. The rods also operated pumps, and the machine system in its general outline is also a monument to the rod-engine and its power transmission principle. The machine demonstrates how the rods' motion was redirected, which was

Figure 10 *Spéléologie minière*: exploring the vertical mines with climbing techniques. The climber is situated outside the remnants of a gallery, evident from the curved forms and remaining parts of the floor. Further down another gallery is visible. The rock outside the upper gallery has been removed in a subsequent phase of mining, probably in the late eighteenth century, when blasting with gunpowder and hoisting with water-wheels made extensive mining of poor ores in side-veins possible, resulting in a number of large extensions in the upper parts of many mines. From the Samuel mine, started in 1630. (BVM-F 320226)

common also at the shaft openings. This was effected by an iron triangle (*Kunstwinkel*).

This water-column engine, like the other engines of the same type running pumps and hoists in Kongsberg, was constructed by the Saxon machine-master Bornemann of Freiberg. This, like other fruitful contacts between Kongsberg and Freiberg, demonstrates the important role of Saxon mining. The engine was constructed in 1877 and the whole arrangement set in operation in 1881. The first modern elevator, a personnel lift,

Figure 11 The *Fahrkunst* in the King's mine, a personnel elevator set in operation in 1881. The two wooden rods moved alternately up and down, and by moving between platforms on the two rods as they simultaneously passed when the platforms came level with each other, the miners made an easy descent and ascent. The machine is demonstrated for visitors to the mine. (Photograph: H. Rock-Löwer/ BVM)

Figure 12 The adit loft in the Christian VII adit, introduced in 1844, divided this main transport and drainage gallery horizontally into two parts. This significantly promoted ventilation and made firesetting in the adit a great success, although it was expected to be very problematic without smoke shafts at distances exceeding 700 m. The adit loft was made partly of wood, partly of brick, as seen here. (BVM 317936)

was installed in the King's mine in 1910. Its successor from 1921, combined with the electric rock and ore lift, is preserved.

Also preserved in this mine is the water-power station, set in operation in 1909 as one of the first underground hydroelectric power-stations in the world (it was preceded by a smaller one in the Samuel mine from 1904). In the King's mine, water from the old dam and aqueduct system was led from the surface with a fall of 333 m to two pelton turbines, one (1907–8) operating an air compressor directly, the other running a generator.

The King's mine exploited the first major ore deposit, found in 1623, and produced about 600,000 kg silver, nearly half the total production of

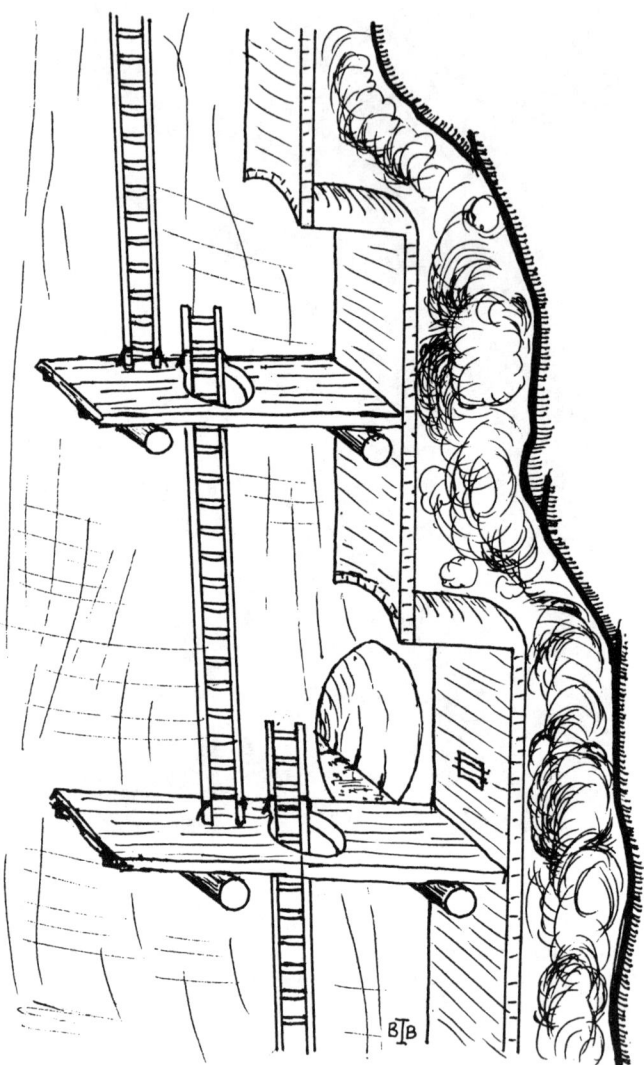

Figure 13 The successful adit loft also prompted a revolution in the ventilation technology in the shaft systems. A high brick wall with arches was a part of the 568 m high smoke channel in the Armen mine, finished in 1867 and still preserved. The success of such installations was responsible for the prolonged use of firesetting up to 1890.

the mines, until it was closed in 1943. At the closing, the director declared the mine preserved as a historical site. For a long time it had already been open to visitors, and the guided visits were taken over by the museum after the closing of the silver mines in 1958.

Not yet included in the guided visits are high stopes and constructions for the ventilation of smoke from firesetting. These constructions are partly brick arches in the adit, parting the lower fresh-air working section from the higher smoke-filled section, and similar high brick walls in the shafts, serving as chimneys (Figures 12 and 13). This ventilation technology, adopted from the 1840s, was a major precondition for the effectiveness and prolonged use of firesetting.

PRESERVATION OF THE SITE: BUILDING COMPLEXES

In the main mining area, all remnants on the surface and underground have been legally preserved from 1992 onwards as a historical site under the Cultural Heritage Act. Except for maintenance of the museum's buildings and the tourist mine with its historic equipment, practical preservation of the site is above all a matter of taking precautions against natural decay of the materials on the surface. One of the most active and expensive tasks of preservation has been the maintenance of that part of the water system which still conducts water. This has until recently been undertaken by the municipality of Kongsberg, for the provision of drinking water. As water is now supplied from groundwater sources, future financing has not yet been decided. Fortunately the maintenance has been done using historically correct materials (rock, peat and wood) and techniques, preserving the authenticity of the dams and aqueducts. Other preservation tasks have been the clearing away of brush and other vegetation from wheel-houses, abandoned aqueducts and mines, partly to counteract damage to the rock walls, partly to make the objects more visible. Historical footpaths have been established and about sixty locations have been equipped with information plates.

At most of the mines no buildings are preserved. At the House of Saxony mine however, a group of buildings dating from the reopening of the mine around 1870 has been preserved: the mining engineer's residence, the miners' dormitory and two shaft buildings. A horse whim has recently been reconstructed. At the mouth of the Christian VII adit, the main communication gallery, there is a larger group of buildings preserved from a period after 1865, among them a magnificent complex of miners' lodgings, and the forge and mechanical driving-belt workshop with nearly complete equipment from the early twentieth century, recently restored to working order.

In Kongsberg town, miners' homes are preserved and also some outstanding buildings from the administrative, educational and religious sphere of mining. The interior of the miners' church from 1761 demonstrates the social stratification and hierarchy of the mining community and the absolute monarchy. Opposite the church is the large wooden building of the Royal *Bergseminarium* of 1786 (Figure 14). This institution was

Figure 14 The wooden building of the Royal *Bergseminarium* from 1786. This institution was founded in 1757 as one of the very first in Europe offering higher technological education, although on a more modest scale than the later famous mining academies of Freiberg, Schemnitz and others. This type of education was transferred to Oslo University in 1814. (BVM-F 312928)

founded in 1757 as one of the first of its kind in Europe, although on a more modest scale than the famous mining academies of Freiberg (Saxony) and Schemnitz (Slovakia), both from the 1760s. Nearby is the administration building of the silver mines and the mining school (for overseers, founded in 1867).

ANCILLARY STRUCTURES

Some ore from the mines was so rich that it could be transferred directly to smelting, some had to be processed in crushing and dressing plants before metallurgical refining. The mines had a central refinery at the waterfalls in the town. The stone building of this refinery, completely renewed in 1844, is preserved together with an ore-storage building and the laboratory, and they now house the museum's collections and administration (Figure 6). In 1922 the refinery was replaced by a new one outside the Christian VII adit; this last refinery is also preserved, and the museum is planning to reinstall the furnaces, which are now in the city refinery, and present exhibitions from the history of metallurgy. Next door to the town refinery and the museum is the Royal Mint site, which in 1686 became the last link in the fully integrated silver production at Kongsberg. The Mint is still in operation, partly within its historical buildings.

INDUSTRIAL ARCHAEOLOGICAL WORK

Historical studies of the silver mines date back to 1631, when a former manager published a report. The first studies in the history of technology in the late nineteenth century were related to evaluations of current company strategy. Firesetting was the theme of the first article of purely historical interest in technology from 1914 (and also of reports in the British *Transactions of the Institution of Mining Engineers* in 1892–3). Later studies evaluated the water-supply system. Now the history of mining technology, in general as well as the operations underground, has become an important research theme in recent years.

The first serious interest in the old mining site at the surface was shown *c*. 1930 by the pharmaceutical chemist Henning Tønsberg, who was also a good photographer and alpinist. He made his way to some of the old mines using rope and ice-axe. He also recorded the surface rock inscriptions.

Since the late 1970s, the ninety or so preserved wheel-houses have been registered, including documentary references to dating and other particularities. Dams and aqueducts and their maintenance techniques have also been registered and documented. In the last few years, the whole complex of objects on the surface has been mapped on scales 1:5000 and 1:20,000, partly on the basis of older maps, partly through field surveys. The objects are registered in the museum's database of sites and monuments of mining in Norway. A great deal of registration and research has still to be done to correct and complete the maps and fill the database with details.

The mines have attracted attention during recent years as new climbing techniques have made these large vertical systems accessible. Since 1985 the author, accompanied by various assistants, has investigated more than 3000 metres vertically and about 30,000 metres horizontally. Many of the galleries are accessible only by using ropes from shafts. The techniques applied and the research performed are largely in the French style of *spéléologie minière*;[13] however, we in Kongsberg started later and have worked with far fewer people and resources than the French, who have been successful in bringing workers from different professions including historians, archaeologists, geologists and engineers into the study of early modern mining and, especially, mining techniques.

In Kongsberg, underground field work has revealed some of the oldest parts of the mines, partly worked by hand with hammer and chisel. Such traces have hitherto been unknown, apparently because the technique was abandoned after the first century of mining. The traces vary in form, but can mainly be seen as 10–30 cm wide fissures at the working faces, possibly representing a preliminary hammer-and-chisel mining of the silver-bearing vein followed by firesetting to achieve trafficable dimensions (Figure 15). In some cases, such sites can be traced in the archive records, thus permitting an exact dating, linking to persons, etc. This was the case with chisel work from 1628, to which year a chisel on the spot also could be dated, as well as the very first adit, completed in 1624 and obviously with firesetting (Figure 1). For the time being, about thirty areas of chisel working have been registered.

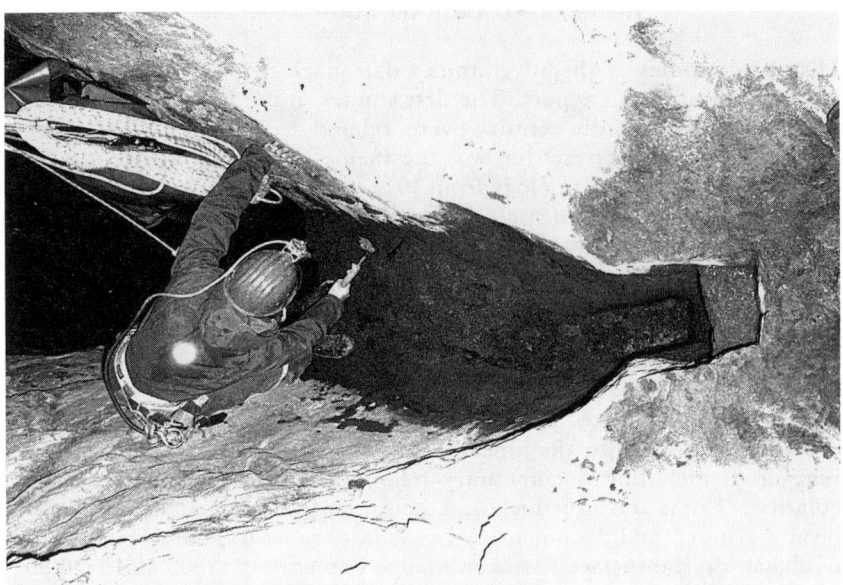

Figure 15 Industrial archaeology: excavation of a shaft wall with a step, seen from above, in the King's mine, possibly from the 1620s. The silver-bearing calcite vein has been followed with hammer and chisel, which is evident from the 15–26 cm wide fissure in the front (it has a maximum depth of 35 cm, one step, and is 115 cm high). This fissure has been followed with firesetting to loosen hard side-rock in order to achieve trafficable dimensions, as is evident from the curved forms behind the fissure, as well as from charcoal and firewood found during the excavation. (BVM-F 317114)

Remnants from other branches of mining technology have also been found. We have inspected the only preserved water-wheel in the mines, 180 m below the surface. It is a reversible wheel for hoisting, with a diameter of 12.5 m and built in 1842–4. Although the wheel consisted mainly of wood, it was in fairly good condition, but the location makes it— for the time being—impossible to maintain.

A dismounted machine of historic interest is a reversible water turbine, also for hoisting, lying with most of its main parts preserved but again in a difficult location, from where we nevertheless hope eventually to recover it. The turbine is of the Saxony *Schwamkrug* construction, and was installed in the Kongsberg mine in 1860. No other machine of this particular type is known to have been preserved.

CONCLUSION

The complexity of the preserved sites and monuments in the overground landscape, formed by nature and culture together, as well as in the underground mine system, constitutes a fascinating field for research and communication in the history of technology. These efforts are facilitated by the mines' archives, providing documentation of what can be observed in the field and in the museum's rich collections, and telling—with the guidance of the historians and pedagogues—the history of Kongsberg as a part of the European mining community. There is still an abundance of material and mines left for examination, as well as for presentation to the public.

Notes and References

1. Production figures for 1623–1805 as well as other quantities, number of workers, productivity etc., are given in B.I. Berg, 'Produktion, Belegschaft und Produktivität beim Kongsberger Silberbergwerk 1623–1805', in Ekkehard Westermann, ed.: *Quantifizierungsprobleme bei der Erforschung der europäischen Montanwirtschaft des 15. bis 18. Jahrhunderts* (St Katharinen, 1988), 127–53.

2. Not included are the greater parts of forestry and overground transportation, which were mainly performed by unemployed labour, e.g. peasants on contracts or as paid duty-work, and involved many persons (roughly around two thousand in the mid-eighteenth century).

3. Detailed archive references will not be given as the issues are fully referenced and documented in a forthcoming book by the author on mining technology in Kongsberg during the period 1623–1914. An illustrated short version is published by the Norwegian Mining Museum and is intended to be translated into English and German.

4. See Graham Hollister-Short, 'Gunpowder and Mining in Sixteenth-and Seventeenth-Century Europe', *History of Technology*, 1985, 10: 31–56; Karl-Heinz Ludwig, 'Die Innovation des bergmännischen Pulversprengens', *Der Anschnitt*, 1986, 3–4: 117–22.

5. See Graham Hollister-Short, 'The Vocabulary of Technology', *History of Technology*, 1977, 125–55. The early history of these machines is outlined in the same author's 'The first half century of the rod-engine (*c.* 1540–*c.* 1600), *Polhem, Tidskrift för Teknikhistoria*, 1991, 3: 192–210.

6. This machine was the first one proposed in 1643, but it was not finished until 1648. The details are given in the monthly accounts, where every machine plant as well as other technical projects are documented throughout the history of the silver mines (in this case: Norwegian National Archives, Kongsberg Silver Mines no. 42.5, accounts 1648, 11th month, Junger Herr mine, payment to the carpenter Georg Schaar of a contracted sum of 190 *riksdaler* for the work). Georg Schaar led the practical building of most of the machines at Kongsberg until 1665. He had migrated from the Harz in 1637 together with his father Peter, also a carpenter. The *Hüttenschreiber* and, from 1646, *Bergmeister* Daniel Barth, who emigrated from Saxony in 1629, was the technical director of the first rod-engines, and also performed the necessary measurements.

7. Christoph Bartels, 'Das Wasserkraft-Netz des historischen Erzbergbaus im Oberharz', *Technikgeschichte* 1988, 3: 182.

8. The major water systems in pre-industrial European mining had the following figures:

District	Dams	Volume (million m^3)	Aqueducts (km)
Upper Harz (Germany)	69	10	198
Schemnitz (Slovakia)	40	7	150
Freiberg (Saxony, Germany)	20	5.7	163
Kongsberg (Norway)	60	1.5	50

See: Rainer Slotta, *Technische Denkmäler in der Bundesrepublik Deutschland 4, Der Metallerzbergbau*, Part I (Bochum, 1983, 210ff); Hugo Haase, *Kunstbauten alter Wasserwirtschaft im Oberharz* (Clausthal-Zellerfeld, 1976); Ján Novák, 'Most Significant Mining Water Systems in Europe until the End of the 18th Century', *Acta historiae rerum naturalium necnon technicarum*, Special Issue 7, Prague, 1974, 39–50; Otfried Wagenbreth and Eberhard Wächtler (eds), *Der Freiberger Bergbau. Technische Denkmale und Geschichte* (Leipzig, 1986); Odd Arne Helleberg, MS and articles in *Langs Lågen*, 1979, also 'Rosentorv', *Volund Norsk Teknisk Museum*, 1981, 77–102.) The figures for Kongsberg (at least) are approximate.

9. Wolfgang von Stromer, 'Wassernot und Wasserkünste im Bergbau des Mittelalters und der frühe Neuzeit', *Der Anschnitt*, 1984, Suppl. 2: 57 gives a probable reference to the machine from Schmölnitz (Gölnitz, Hungary) from 1439. The reversible wheel was until the mid-sixteenth century obviously used mainly for hoisting water.

10. The early history and development of the rod-engine is a theme of Graham Hollister-Short, *op. cit.* (5). The first use of the term *Stangenkunst* (for a rod-engine built in 1551) occurs in Johannes Mathesius's *Chronicle of Joachimsthal (Jáchymov)*, Bohemia. The first known picture of field rods is from Jean Errard of Bar-le-Duc' work of 1584. See also Stromer, *op. cit.* (9), 63.

11. Sten Lindroth, *Christopher Polhem och Stora Kopparberget: Ett bidrag till Bergsmekanikens historie* (Uppsala, 1951), 38ff.

12. Albert Riechers, *Erfindungen im Harzer Erzbergbau* (Clausthal-Zellerfeld, 1975), 5ff.

13. An outstanding French study is: Bruno Ancel and Pierre Fluck, *Une exploitation minière du XVIe s. dans les Vosges. Le filon Saint-Louis du Neuenberg (Haut-Rhin). Caractères et évolution*, Documents d'archéologie française No. 16 (Paris, 1988); see also outline articles in *Archéologie Médiévale en Alsace*, 1987, 3: 105ff.

Does the Development of Mobility in Traffic Follow a Pattern?

HERMANN KNOFLACHER

It might seem obvious that movements taking place in traffic systems display regularities or patterns. On the assumption that traffic flows do in fact exhibit regularities, it follows that if we were able to detect and decode such patterns we could then go on to modify them by changing the factors that produce them.

EXISTING TRENDS: THE PREVAILING VIEW

The increasing rate of motorization has caused a decrease in the patronage of public transport systems, as shown by Voigt[1] (Figure 1). This tendency may be observed in the existing data from all countries regardless of social,

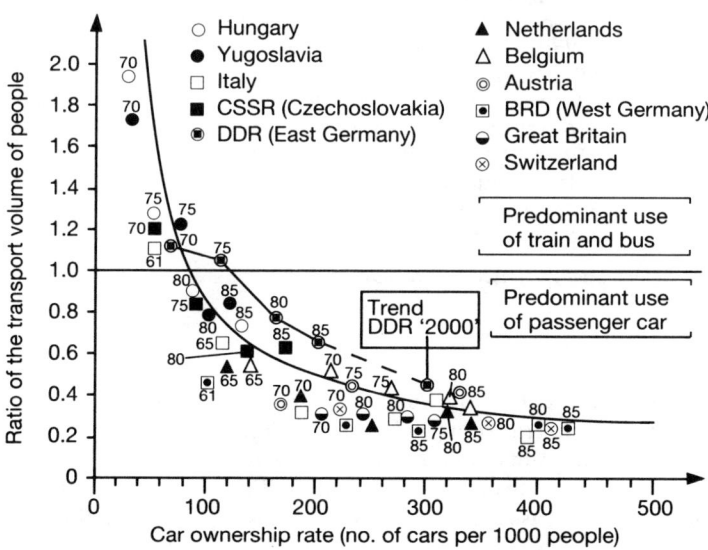

Figure 1 Ratio of public to private transport volume against car ownership rate.
Source: Note 1, p. 109.

125

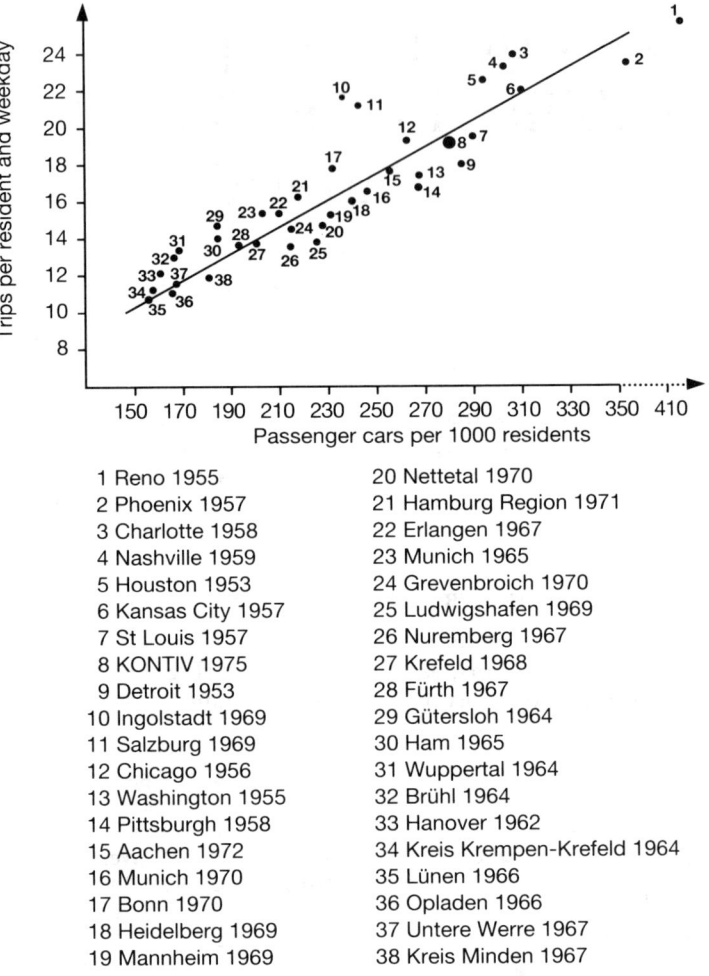

1 Reno 1955
2 Phoenix 1957
3 Charlotte 1958
4 Nashville 1959
5 Houston 1953
6 Kansas City 1957
7 St Louis 1957
8 KONTIV 1975
9 Detroit 1953
10 Ingolstadt 1969
11 Salzburg 1969
12 Chicago 1956
13 Washington 1955
14 Pittsburgh 1958
15 Aachen 1972
16 Munich 1970
17 Bonn 1970
18 Heidelberg 1969
19 Mannheim 1969

20 Nettetal 1970
21 Hamburg Region 1971
22 Erlangen 1967
23 Munich 1965
24 Grevenbroich 1970
25 Ludwigshafen 1969
26 Nuremberg 1967
27 Krefeld 1968
28 Fürth 1967
29 Gütersloh 1964
30 Ham 1965
31 Wuppertal 1964
32 Brühl 1964
33 Hanover 1962
34 Kreis Krempen-Krefeld 1964
35 Lünen 1966
36 Opladen 1966
37 Untere Werre 1967
38 Kreis Minden 1967

Figure 2 Relationship between average weekday trips and car ownership rate in thirty-eight cities.

Source: Note 2, p. 79.

economic and political conditions. No country has been able to reverse this trend, even if their transport policy has sought to do so.

It seems that mobility patterns follow a kind of natural law, being tied to the increase in motorization. Current traffic policy and traffic planning are based on the assumption that 'increasing motorization increases mobility'[2] (Figure 2). This is a basic hypothesis of traffic policy planning, valid until proven otherwise.

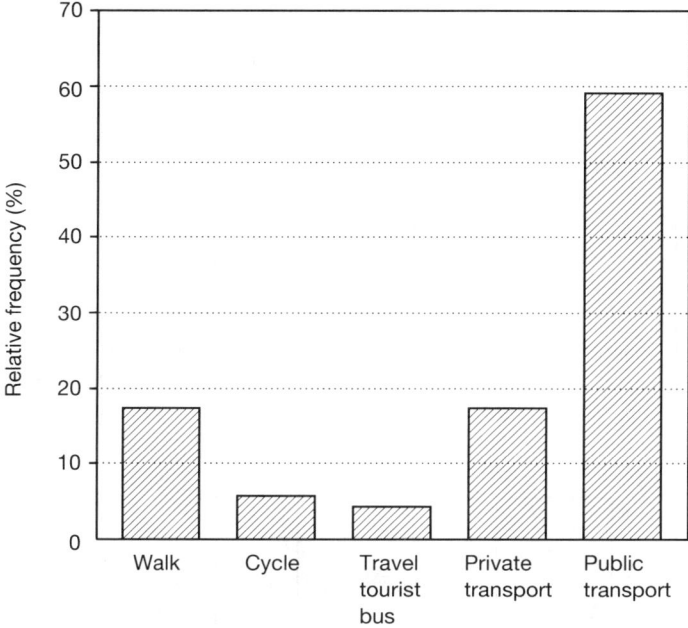

Figure 3 Means of transport to reach the first district of Vienna.

Source: Note 3.

Examination of the hypothesis

Modal split of the type of transport used by shoppers in Vienna's city centre
Figure 3 shows the result of a survey made in the city centre of Vienna.[3]
Vienna has about 380 cars per 1000 inhabitants, but we can see that only
15 per cent of customers use the car for shopping in the city. The majority
use public transport or walk. Vienna has an excellent public transport
system in the city centre in the form of its Metro and a very pleasant
pedestrian zone. We can see, therefore, that the hypothesis is not valid for
customers in the city centre of Vienna.

Modal split of commuters dependent on parking places
Only 10 per cent of car-owning commuters who have a reserved parking
place at their destination in the city use public transport (Figure 4).
However, if they do not have a reserved parking space, about 30 per cent
use public transport regularly.[4] (The parking policy in Vienna is not very
strict, so it is not too difficult to find a parking place.) This is the second
observation contradicting the hypothesis.

Effect of improved supply of public transport on modal choice
Traffic flow on radial roads to Vienna has increased at the same rate, or
faster, than on the average Austrian road. In the 1970s, a *Schnellbahn* system
(trains with a convenient half- or quarter-hour interval) was introduced on

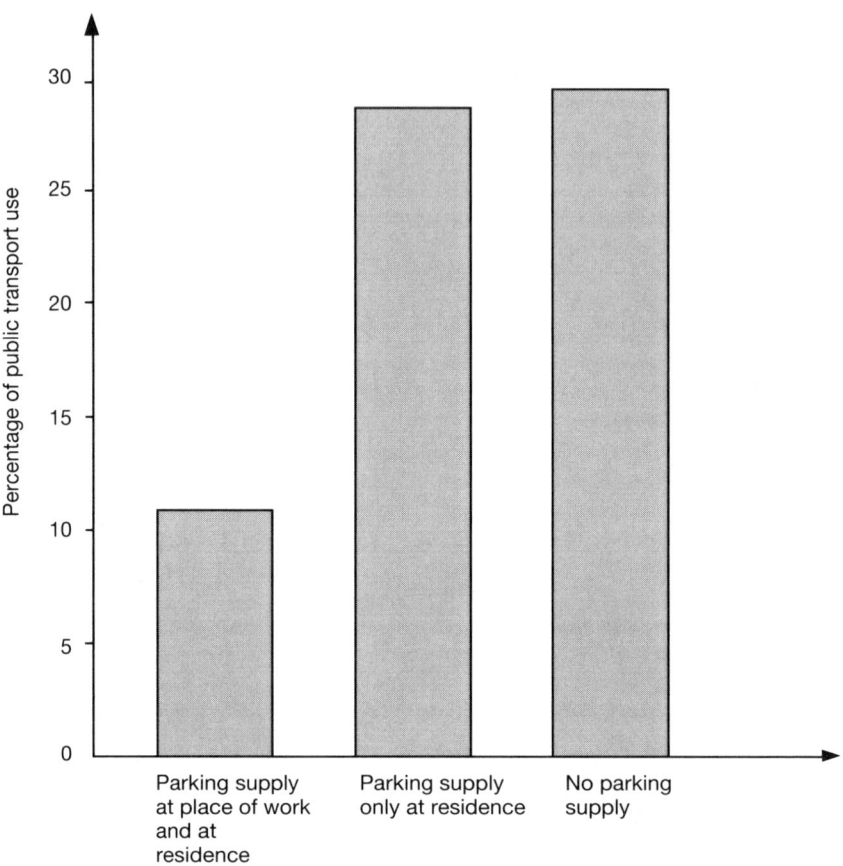

Figure 4 Participation of the season-ticket user.

Source: Note 4.

lines parallel to some of these radial roads, so that commuters would have
an alternative to driving. The effect on road traffic was clearly visible.[5]
Traffic flow at the end of the 1980s was no greater than at the beginning
of the 1970s, although the total Austrian traffic growth rate continued
to increase, as seen in Figure 5. This is the third observation contradicting
the given hypothesis.

We can conclude from these three observations that the infrastructure
and the quality of the traffic system both strongly affect the modal split,
the possibility in this case of going either by rail or by road, and thereby
also affect mobility patterns.

THE QUESTION OF MOBILITY

Car mobility is only one aspect of the physical mobility of people. Figure
6 shows a cross-section of the mobility of people between the ages of 6 and

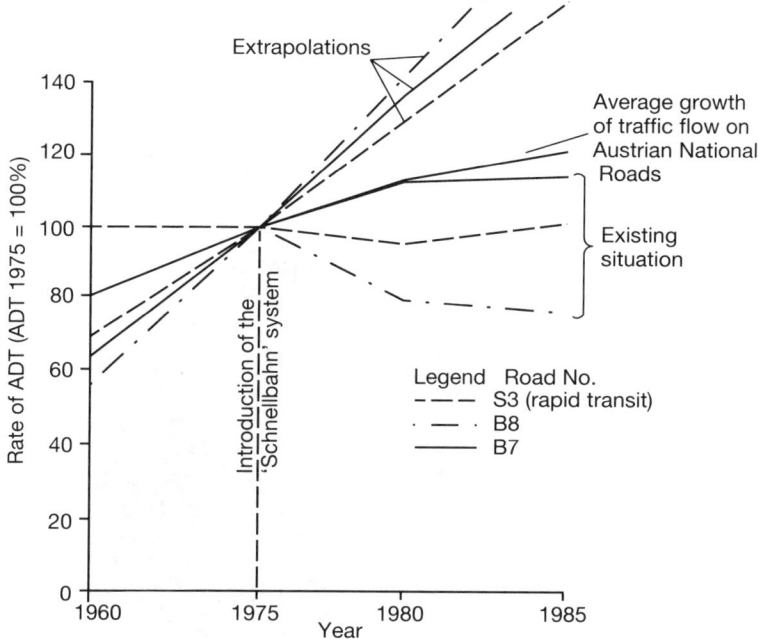

Figure 5 Development of traffic growth with respect to the rapid transit (Schnellbahn).

Source: Note 5.

90 living in a typical medium-sized Austrian city.[6] The absolute number of trips per person depends on age.

Car traffic constitutes only a part of total mobility, the bulk of which is pedestrian mobility, bicycles and public transport. It is important to note that mobility denotes not only physical but mental, social and housing mobility as well. These three factors play a large role in physical mobility because all of them can either inhibit or enhance physical mobility. Therefore it is useful to define mobility in terms of one person and his or her number of trips per day instead of in terms of transport modes. Studies worldwide show that this so-called mobility is not dependent on the development of technical traffic systems alone. Social demographic changes are responsible for most of the increasing mobility rate of recent decades. One of the reasons is the decreasing number of people per household. For instance, people living alone will make more external trips than people living as part of a couple, simply because the single person has to find social contact outside the domestic environment.

We have to accept that the number of trips per person per day does not change much if the transport modes are changed. Several studies show that this mobility rate seems to be constant. The distance travelled, not the number of trips, changes. Average travel time has remained constant, but increasing traffic mobility has caused the trip distance to increase.

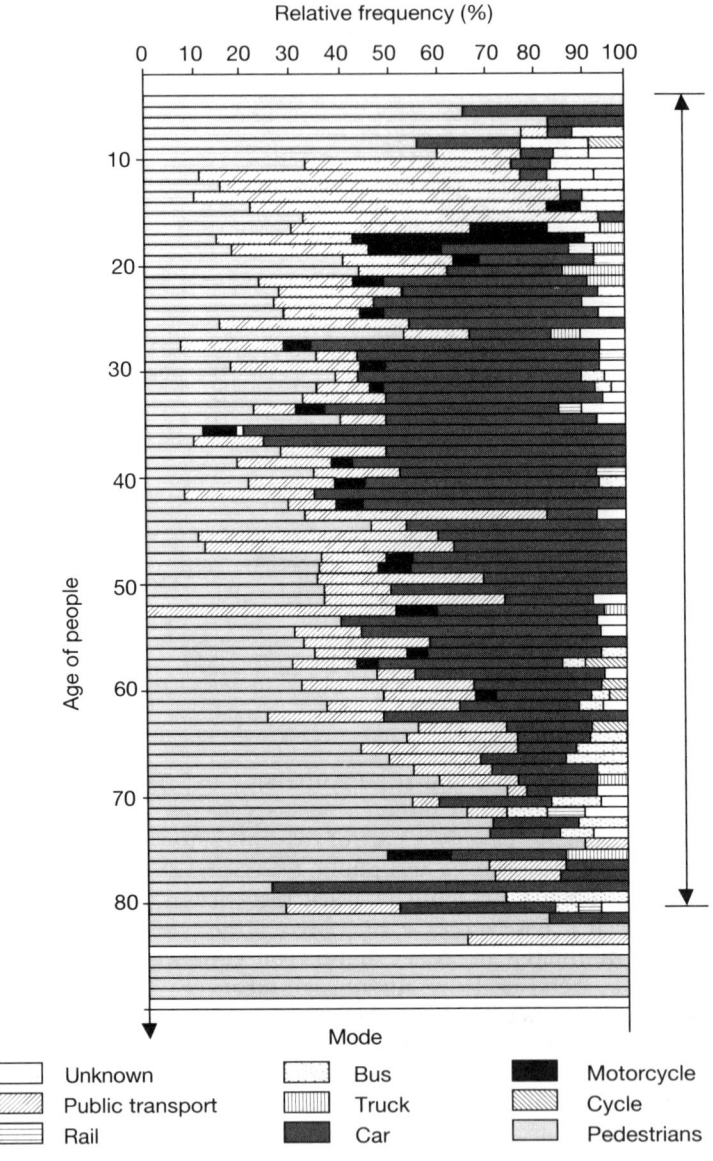

Figure 6 Distribution of mobility among different age groups and modes of transport.

Source: Note 6.

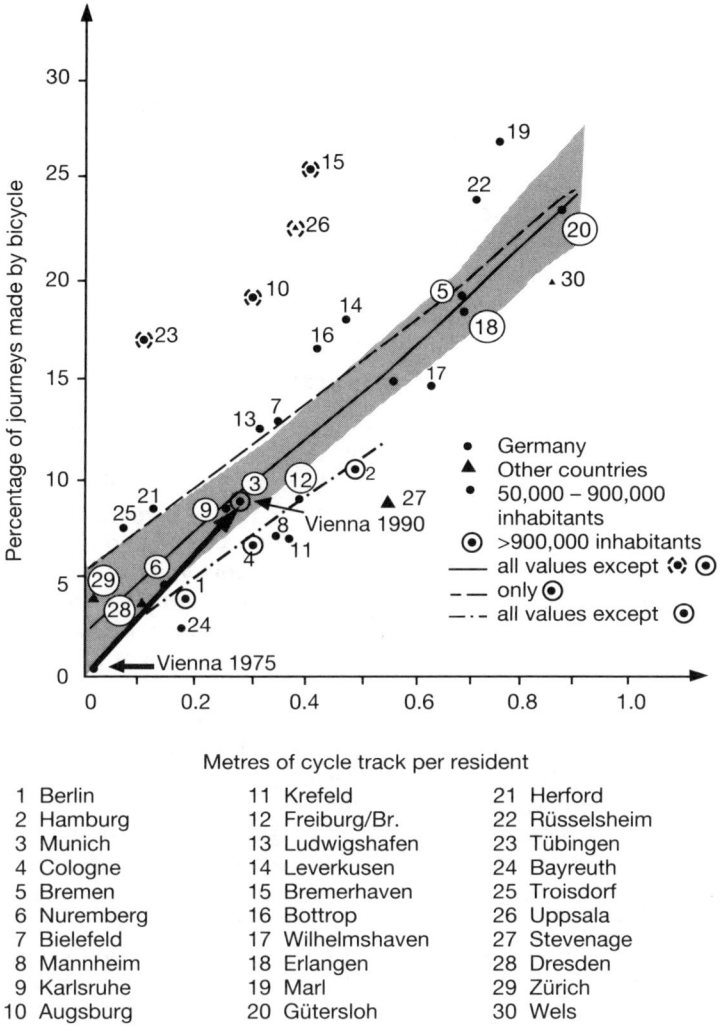

Figure 7 Relationship between choice of transport mode (cycle) and metres of cycle track per resident.

Source: note 7, p. 115.

Therefore increasing traffic-system speeds will not result in time savings. We cannot understand or change traffic systems by calculating time savings, since, as far as individual mobility is concerned, there is no time saving in the transport system.

What kind of regularities influence the modal split?

In 1975 a survey of cycle-tracks and the number of cyclists was carried out in German cities.[7] The results are shown in Figure 7. An increase in the number of cycle-tracks leads to an increase in the number of cyclists. This figure shows a relationship between human behaviour and infrastructure, in this case cycle-tracks. From this diagram we can see that Figure 2 shows an apparently incomplete correlation, one which presupposes the existence of a road network. For people to make car trips, roads must exist. If no road network exists, no corresponding pattern of car mobility will appear, even though people have cars.

If we look at the limited public space in cities and compare it to the basic space demands of different kinds of travel modes, it becomes obvious that we must set certain priorities. How can we move human behaviour in the direction of these priorities?

HUMAN BEHAVIOUR AS A RESULT OF EVOLUTION

Road and city planning are designed in terms of time and distance scales. If planning measures do not produce the human behaviour expected by planners and traffic politicians, then these two groups blame irrational human behaviour. It is unlikely, however, that the majority of people behave irrationally. Each person behaves in a rational and logical way as he or she sees the situation. Possibly this behaviour might be mistaken, or have negative effects in relation to the needs of the whole system, but not to the existing situation as the person sees it. Therefore we have to accept that people are not irrational but that the assumptions of politicians and planners are unwarranted.

Since we know that the traffic system does not follow any simple calculation of time-saving, the basis of traffic planning and traffic economy, we now have to question whether the traveller's perception of time corresponds to actual time measured by the watch.

In 1972, Walther conducted an interesting study in Bielefeld.[8] He asked public-transport users to estimate their walking time compared to their whole travel time. Then he compared their estimates with the real walking time. If the estimated time for walking was the same as the entire travel time then people would judge time accurately. One minute travelling in a bus would equal one minute's walking; if we divided the estimated time by the measured time, we would always get 1. However, the actual observations were quite different,[9] as shown in Figure 8. With increasing walking distance, an overestimation of walking time occurred when compared to the whole travel time. This result was called the 'time value factor'. The reciprocal of the time value factor may be defined as 'attractivity' (Figure 9). The attractivity of a walking distance declines sharply with increasing distance. A walkway 300 metres long has only about 20 per cent of the attractivity of one 50 metres long. We repeated these studies in Vienna[10] and the results supported those of Walther.

Bees follow the same pattern

Karl von Frisch observed that the frequency of bees' dances used to give distance information decreases with increasing distance to the feeding place.[11] We see that Frisch's observations resulted in an exponential curve similar to Walther's (Figure 10i).

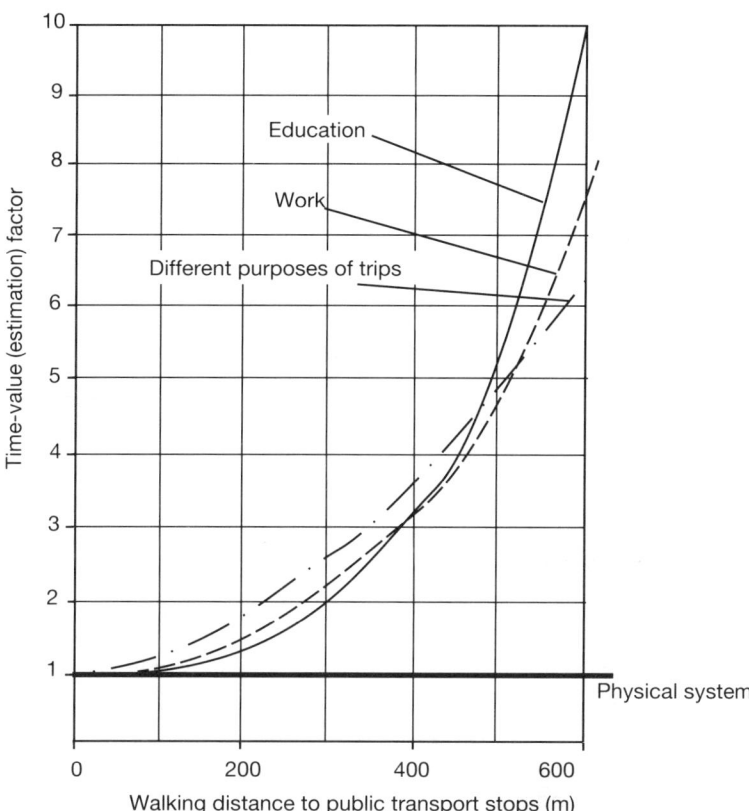

Figure 8 Time estimation of walkways compared to public transport travel times.
Source: Note 8, Figure 42.

What is the reason for this similarity?

I think I can explain the similarity by means of the following analysis. In 1956 Karl von Frisch conducted an interesting experiment with bees that can also be taken as a key experiment for traffic planning.[12] Bees have the possibility of a modal split since they can both walk and fly.

The experiment modified the bees' environment by means of a channel leading from the beehive to the feeding place. The bees were forced to walk since they could not fly. Frisch and his colleagues recorded the information the bees gave after returning home when the channel was lengthened. It was interesting to find that after the channel was extended to lengths between 3 and 4 metres, the bees informed their fellow workers at home about a feeding place some 80 metres (flight distance) away. This means that the bees were 'lying', or do not measure distance as we do. Frisch discovered that for both distances, 4 metres walking or 80 metres flying, the energy consumption was the same. This was the cue to compare the energy consumption of man when walking, driving or using public transport.

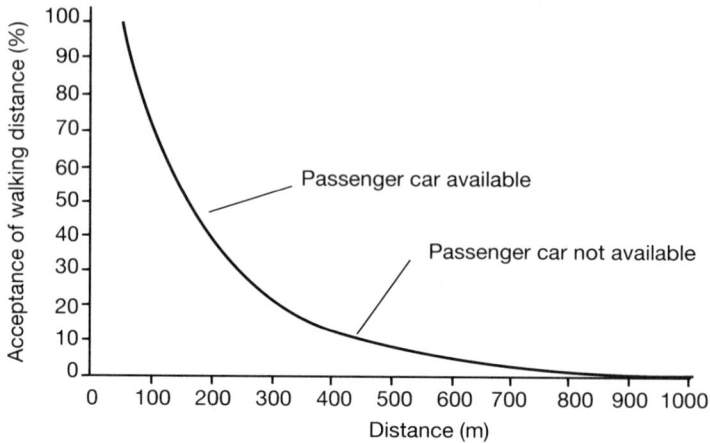

Figure 9 Acceptance of walkways as a function of distance.

Source: Note 8, Figure 23.

HOW DO WE CALCULATE ENERGY CONSUMPTION?

Frisch found a correct mathematical description or function (Figure 10ii), but I believe he drew the wrong conclusion.

At the beginning of the nineteenth century, Weber[13] and Fechner[14] discovered a fundamental law governing the reactions of organisms to external irritations. This law is called the Weber–Fechner sensation law; sensation (s) is equal to the natural logarithm of intensity of irritation (I).

$$s = \ln I$$

In the light of all this—what is traffic planning? Traffic planning consists of nothing more than the changing of external irritations and, therefore, sensations. If we were to use human body energy demand to describe traffic behaviour, then we should be able to find a way to explain and modify human traffic behaviour.

The sensation of a daily walk to work, shops or other interesting things is perceived as resistance. Resistance is recognized as negative. If we introduce a negative sensation into the Weber–Fechner law, we obtain the negative exponential (e) functions, which can be found in most functions describing the resistance laws of traffic systems. However, this explanation gives us a much deeper insight into the problem when compared to a formal description. We can now explain the traffic resistance law on the basis of the 'inner mechanism of man'.

The function is twofold. External irritation can also create positive sensations. Therefore it was interesting to check whether this twofold function

Figure 10(i) Dances of bees as source of distance information.

Source: Note 11, p. 61.

could be observed empirically. A student was given the task of observing the differences between public transport users going through a normal environment (car-oriented), and through a human-oriented environment (walking through parks or pedestrian zones) to reach their public transport stop.[15] The conditions were defined as

(a) an unattractive environment, car-oriented.

(b) an attractive environment, pedestrian zones and parks.

The results are shown in Figure 11. The curves are parallel, but the attractive environment 'extends' the attractivity of a walkway by up to 70 per cent or more.[16] In an unattractive environment, people tend much more strongly to avoid walking and to use a car if possible.

PEOPLE ARE 'CAPTURED' BY THE CAR AS A RESULT OF THEIR OWN ATTITUDE

Balances of energy and irritation are probably taking place in the brain at very deep evolutionary levels, presumably far below the level of consciousness.

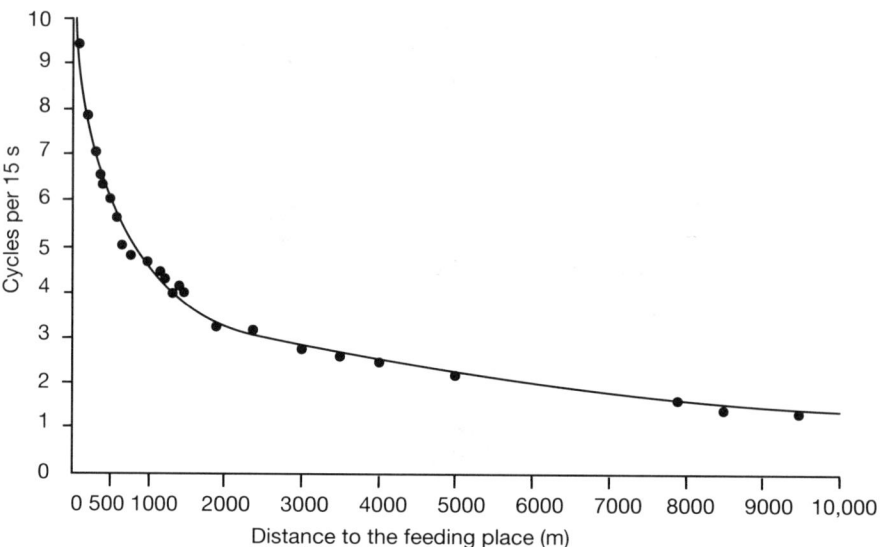

Figure 10(ii) Relationship between dancing speed and distance to feeding place.
Source: Note 11, p. 70.

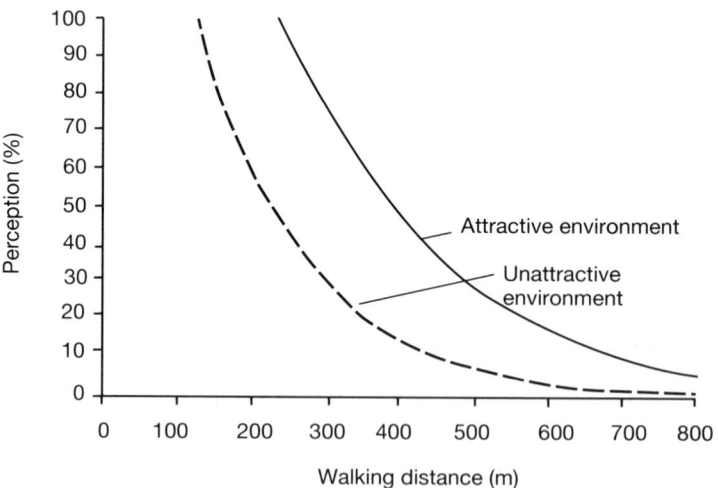

Figure 11 Influence of municipal structure on perception of travelling time: traffic
to place of work, free alternative means of communication.

Source: Note 10, Figure 20.

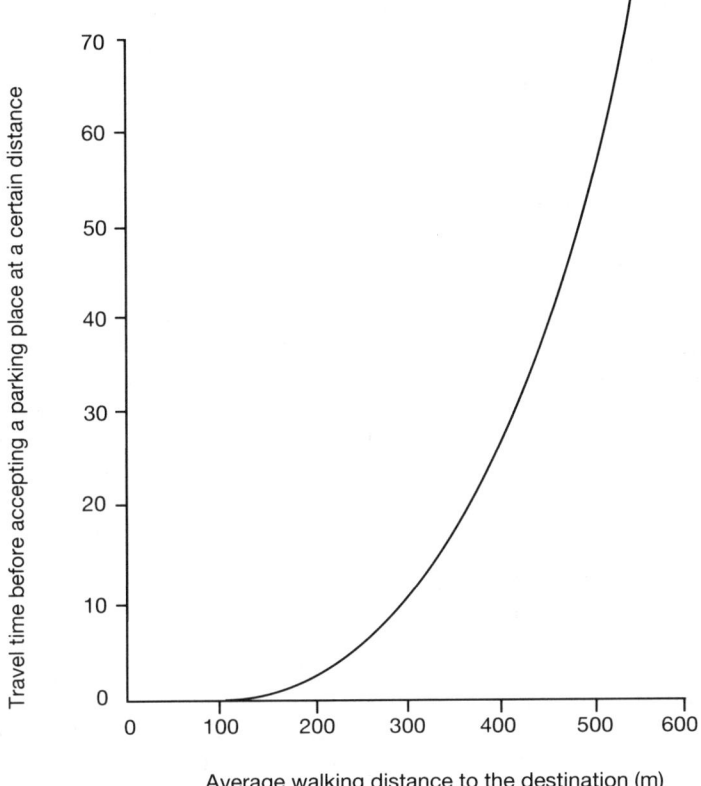

Figure 12 Average search time to find parking place closer to destination before accepting a certain walking distance.

Source: Note 17, p. 111.

It was also interesting to observe, therefore, for how long car users would search for parking places before they accepted a particular walking distance. This task was given to another student as a subject for his thesis.[17] The results are shown in Figure 12. We see that a distance of 100 metres or less between the destination and the parking place is accepted immediately by the car drivers. If the distance is increased to 300 metres, the car drivers look for a parking place closer to the destination for an average of 12 minutes; and if the distance is 400 metres, the searching time for a closer parking place is 28 minutes. This function, as well as the attractivity function in Figure 11, can be very easily calculated in terms of the Weber–Fechner law and energy balances.

We are captured by the construction of our cars for many reasons. The body energy consumption for sitting is much lower than for the upright standing position. The 'power of the legs' is increased by about 600 to 700 times if we use a car compared to walking, but the brain has not grown at the same rate. The brain concludes that we need less body energy to drive than to walk.

As car drivers, we are leaving human society behind since we are leaving the realm of the two-legged. Car drivers are 'four-legged' as they move. It is as if the space of activity has become similar to sitting on a branch in a tree, having a 'steering branch' in one hand, a 'gear branch' in the other hand, the legs having a 'braking branch', 'accelerating branch' and a 'clutch-pedal branch', and the whole tree is moving—really a fascinating thing.

The car is allied with the subconscious levels of our brain. Our captivity is nearly complete.

Consequences for our question

In all countries of the world we observe the same basic mistakes in the organization of traffic systems. The control mechanisms are not working efficiently, or do not exist at all. Traffic policy and traffic planning are oriented towards the big elements like roads, computerized signalling, railways, etc., not towards the small elements, the cells of the traffic system. We can define the cell of the traffic system as the household or the individual.

We can now look at the individual and his position *vis-à-vis* different traffic modes. Today, the situation worldwide is the same. The car is parked in front of the house or in the garage. The walking distance to the parked car, therefore, has nearly 100 per cent attractivity. The public transport stops are far away, sometimes 700 metres or more, thus the remaining attractivity of public transport is only a few per cent. Under these circumstances public transport has no chance at all. With increasing motorization, the level of public transport use must therefore decrease, following a negative e-function which is the result of the prevailing traffic organization and the relationship of cars to human activities compared to public transport stops.

What Voigt[18] has observed is nothing more than the realization of the Weber–Fechner law in this man-made environment. Therefore, the key for changing the system is not the planning of 'big elements' but much more a proper organization of the 'small elements'. This means the proper organization of parked cars, and caring for real human behaviour, which is oriented towards maximizing positive irritation (convenience) and minimizing negative irritation (consumption of body energy).

SOLUTION OF THE PROBLEMS

If we accept man as a reality, and not as a creation of planners and politicians, then we have to respect his fundamental nature. This means that we have to take into account his real behaviour, not his desired behaviour. If traffic policy demands some changes in public transport (most political programmes are promising priority for public transport), then this takes on real meaning. The consequences of this political goal is that we have to introduce the same distance between all human activities (living,

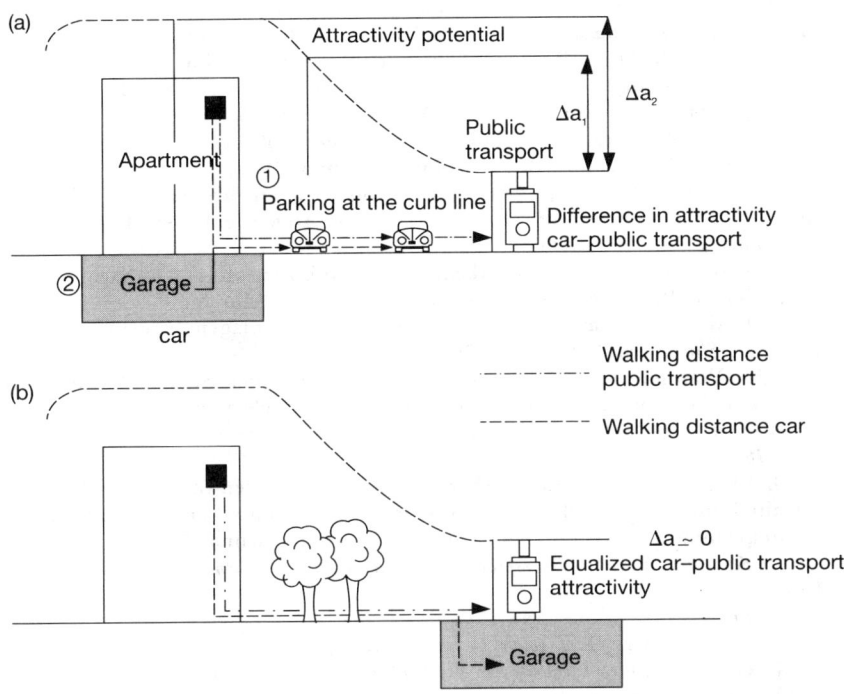

Figure 13 Solution of the problems: (a) existing situation and (b) proposed system.
Source: Note 19, pp. 150–4.

working, leisure, shopping, etc.) and the parked car that exists between these activities and the public transport stop[19] (Figure 13). This means that our residential areas must be cleared of cars. The cars must be stored in central garages at least as far away as the next public transport stop. The result would be car-free pedestrian environments with possibilities for cycling, social contacts, leisure, etc. Obviously such an arrangement would involve a massive change in our society and our economy, but the results would be extremely positive: more stable, well-developed residential areas, which also fulfil ecological needs. This kind of organization would take care of human needs as they are and not how we would like them to be. They present a tremendous challenge for all policies and strategies, and we cannot avoid either the needs or the implementation of these kinds of solutions if we accept the real nature of man on the one hand and wish to pursue a serious traffic policy on the other.

Notes and References

1. W. Voigt, 'Ausgewählte Aspekte zur Situation im Verkehrswesen der DDR', *Strassenverkehrstechnik*, 1990, 3: 104–12.

2. H. Hautzinger and P. Kessel, 'Mobilität im Personenverkehr', *Forschung*

Strassenbau und Strassenverkehrstechnik. Forschungsberichte herausgegeben vom Bundesminister für Verkehr, Abteilung Strassenbau, No. 231, Bonn–Bad Godesberg, 1977.

3. H. Knoflacher, 'Parkraumbedarf der Wirtschaft im 1. Wiener Gemeindebezirk.' Studie durchgeführt im Auftrag der Wiener Handelskammer—Verkehrspolitische Abteilung, Vienna, August 1990, p. 35.

4. H. Knoflacher, 'Verkehrskonzept Klosterneuburg', Durchgeführt im Auftrag der Stadtgemeinde Klosterneuburg, Vienna, 1987.

5. G. Steierwald *et al.*, 'Automatische Strassenverkehrszählung—Jahresbericht 1985', Bundesministerium für wirtschaftliche Angelegenheiten, Bundesstrassenverwaltung, Vienna, 1983.

6. H. Knoflacher, 'Generalverkehrsplan Ried im Innkreis', Durchgeführt im Auftrag der Stadtgemeinde Ried im Innkreis, Vienna, March 1983.

7. H. Knoflacher and H.P. Kloss, 'Radverkehrsanlagen: Ergebnisse einer Erhebung', *Strassenverkehrstechnik*, 1979, No. 4: 109–16.

8. K. Walther, 'Nachfrageorientierte Bewertung der Streckenführung im öffentlichen Personennahverkehr', Dissertation, vorgelegt an der Rheinisch-Westfälischen Technischen Hochschule Aachen, Frankfurt an der Oder, 1973.

9. *Ibid.*

10. O. Peperna, 'Die Einzugsbereiche von Haltestellen öffentlicher Nahverkehrsmittel im Strassenbahn- und Busverkehr', Diplomarbeit am Institut für Verkehrsplanung der Technischen Universität Wien, Vienna, May 1982.

11. Karl von Frisch, *Tanzsprache und Orientierung der Bienen* (Springer Verlag, 1965).

12. *Ibid.*

13. E.H. Weber, *De Tractatu* (Leipzig, 1834).

14. G.T. Fechner, *Elemente der Psychophysik* (Leipzig, 1860).

15. Peperna, *op. cit.* (10).

16. *Ibid.*

17. Th. Macoun, 'Zugangszeiten zu Parkplätzen', Diplomarbeit, ausgeführt am Institut für Verkehrsplanung der TU-Wien, September 1984.

18. Voigt, *op. cit.* (1).

19. H. Knoflacher, 'Zur Frage des Modal Split', *Strassenverkehrstechnik*, 1981, No. 5.

The Greeks and the
Early Windmill

MICHAEL J.T. LEWIS

ABSTRACT

This paper attempts to evaluate the contribution of the Greek world to the
early development of the windmill. The inspiration behind Hero of Alexandria's vertically sailed toy of the first century AD, it is suggested, was a
rotary fan. The horizontal windmills of Persia, recorded from about AD
800, find their closest counterparts both in time and place in the Byzantine
Empire, prompting the suspicion that they came west in the wake of the
Seljuq invasion of Asia Minor in the 1070s. Here, in the twelfth century,
according to the model proposed, the horizontal mill was also married to
Hero's wind-wheel to beget the first vertical windmill, the paltrok. Before
long, this was transmitted north into Russia and, by the Crusaders, west
to France, Flanders and England whence, before it was adapted into the
post mill, it was also introduced to Portugal. The more sophisticated tower
mill seems equally to be of Aegean origin, invented and carried westwards
along the Mediterranean and into France during the first half of the
thirteenth century. The Byzantine world, it is argued, thus played the
pivotal role in developing and diffusing one of the most fundamental of
medieval machines.

INTRODUCTION

The early history of the windmill has engendered much debate. Two basic
facts are not in doubt. One is that the first categorical evidence for windmills
comes in the tenth century, with a probable reference about 800 and a
possible hint in 644. These mills were in Sistan on the borders of the present
Iran and Afghanistan where they have remained to the present day; and
they were horizontal mills, with the sails mounted on a vertical axle that
turned the millstone without gearing. The second fact concerns the vertical
windmill, essentially of the kind still with us in the West, which, with its
angled sails and geared drive, marks a very different approach. This is first
recorded in western Europe, in the triangle formed by east and south
England, Normandy and Flanders, around 1180. What was the link (if any)
between these two regions, half a world apart? That is the nub of the
problem. The old notion that the Crusaders met the windmill in the East
has long been abandoned, on the grounds that no Crusader went near
Sistan, and that there is no evidence whatever that the Persian windmill

141

was imitated by Muslims further west.[1] A few support the concept of independent invention; Jespersen argues an interesting scheme which will call for comment later; but most now favour the theory of stimulus diffusion. Crusaders in the Holy Land, they say, heard of the Persian windmill at second or third hand from Saracens who had themselves seen it. All they took back home was the bare concept of a mill powered by wind, which they interpreted in their own very different fashion: first the post mill, later the tower mill.[2]

In this scenario the lands of the eastern Mediterranean play no obvious role. The Crusaders are supposed to have introduced the western windmill—presumably the post mill—to Palestine. Otherwise, it is thought that the Mediterranean in general was innocent of the post mill and knew only the tower mill. The earliest certain references to windmills here are given as 1332 (Italy) and somewhere between 1249 and 1389 (Greece).[3] The general feeling is that the Mediterranean mill was an import from north-western Europe; even Jespersen, who is almost or quite alone in seeing it as marking an intermediate stage between the Middle East and Europe, does not argue the case in detail. Only one article, thin and unsatisfactory, has hitherto tried to investigate the early history of the Greek windmill,[4] and this paper attempts to fill the gap; or rather, because there is much work yet to be done, to start to fill it.

I offer an explanation of Hero of Alexandria's enigmatic little windmill, and some new information about medieval Greek mills, which I attempt to put in perspective. Since—to anticipate my conclusions—I agree with Jespersen's implication that the Byzantine world was the melting pot where the eastern horizontal mill was recast as the western vertical mill, we have first to look at what had gone before. At the other end of the time-scale, I set my finishing line at about 1500. In between, I envisage various stages of windmill development which I try to set, however briefly, in the technical and cultural context of their time and place. And while I have delineated with a broad brush the westward and northward diffusion of the windmill, I leave it to those with more expertise than I have to fill in the details here, and indeed to test the thesis, both on its fringes and in its core, against their own knowledge.

There is much to test. While research on windmills of later centuries continues unabated, work on their origins is to some degree stuck in a rut. Hard evidence being elusive, I have proposed a model which fits the recorded facts, offers an explanation of how they are connected, and identifies areas where further research is needed. The result is necessarily no more than a theory, though I hope a plausible one. Formulating models of this kind to be proved, disproved or amended seems the best route towards understanding the evolutionary mysteries of one of the most fundamental of medieval inventions; and I unashamedly hope to stir the debate— and research—back into life.

HERO'S WINDMILL

Hovering uncertainly in the prehistory of the windmill is a curious toy described by Hero of Alexandria: a windmill undoubtedly it was, but nobody seems to know quite what to make of it. Hero's date is no longer the mystery it once was: he flourished in the second half of the first century AD.[5] The diagram which accompanies the relevant chapter has been rejected as a later addition by Christian or Muslim scribes who knew the windmill proper,[6] and this dismissive view is still being repeated[7] despite Drachmann's proof that the diagram is genuine.[8] No fewer than 94 manuscripts of the *Pneumatica* are known and all contain the diagram in question. Drachmann rightly argues that since the text has reference letters to a diagram, both are of the same age; that the archetype of all our manuscripts carried both; and that it is earlier than about AD 550. At that date there were no muslim scribes to interpolate it and no 'windmills proper' to inspire it. There is no reason whatever to doubt that text and diagram go back to the first century AD.

The diagram copied here (Figure 1) is from the oldest and best manuscript.[9] The text reads:[10]

> Construction of an organ, so that when the wind blows it produces the sound of a pipe.
>
> Let α be pipes, and $\beta\gamma$ be the transverse tube connecting with them, and $\delta\epsilon$ the upright tube, and $\epsilon\zeta$ another transverse tube leading from it to a cylinder $\eta\theta$ which has its inner surface made smooth for a piston. Into this let the piston $\kappa\lambda$ be fitted, which can pass easily into it; to this let there be fastened the rod $\mu\nu$ which engages another rod $\nu\xi$ pivoting about the axle $\rho\pi$; and at ν let there be a loose-fitting pin; at ξ let the small plate ξo be fitted, and next to ξo let there be an axle σ, and let it turn on iron pivots in a frame that can be moved about. Let there be fastened to the axle σ two discs υ, ϕ, of which let υ have pegs engaging the plate ξo; and let ϕ have blades like the so-called *anemouria*. So when they are struck by the wind they are all driven and turn the disc ϕ, and the axle is turned and hence the disc υ, and the pegs on it hitting the plate ξo at intervals will lift the piston $\kappa\lambda$; and as the peg moves on, the piston will fall and will force out the air in the cylinder $\eta\theta$ into the tubes and pipes and will make the sound. It is always possible to turn the frame carrying the axle towards the wind that is blowing so that the rotation becomes stronger and more continuous.

All this is basically clear enough. The diagram has suffered less in transmission than have many of Hero's, and Schmidt's reconstruction (Figure 2) seems adequate in its essentials, even if his air pump seems unduly large. Although neither text nor diagram indicates that the blades are angled, he inevitably makes them so. The only alternative would be for the blades to be set parallel to the axis and struck tangentially by the wind; but Hero says nothing about shrouding the bottom half, as such an arrangement demands. What he does specify to hold the axle is a *pegma*—an open framework, not a box or shield—which, since he fails to draw it, Schmidt supplies. Hero gives no details of the organ itself because he has described it fully in the previous chapter, although one does miss here the necessary wind reservoir. The standard Graeco-Roman organ, invented by Ctesibius at Alexandria in the third century BC, had an air pump on either

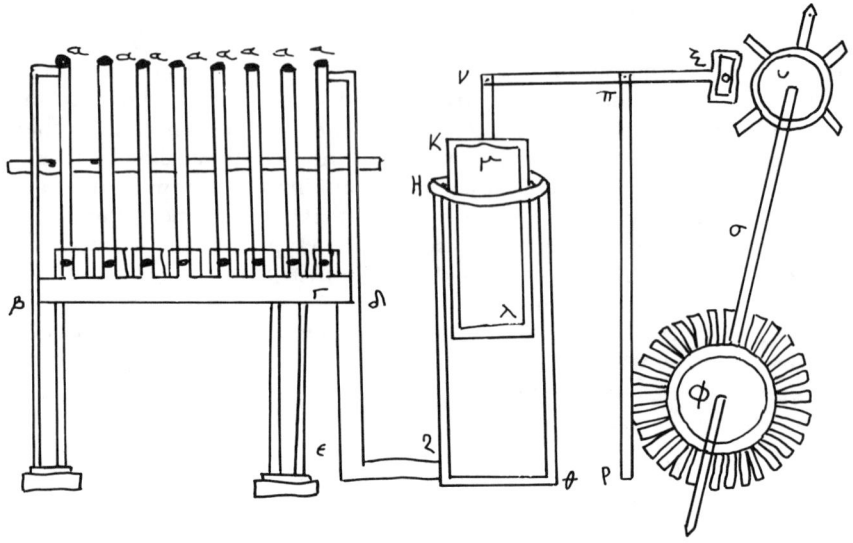

Figure 1 Hero's diagram of wind-powered organ (after Venice, Marcianus 516).

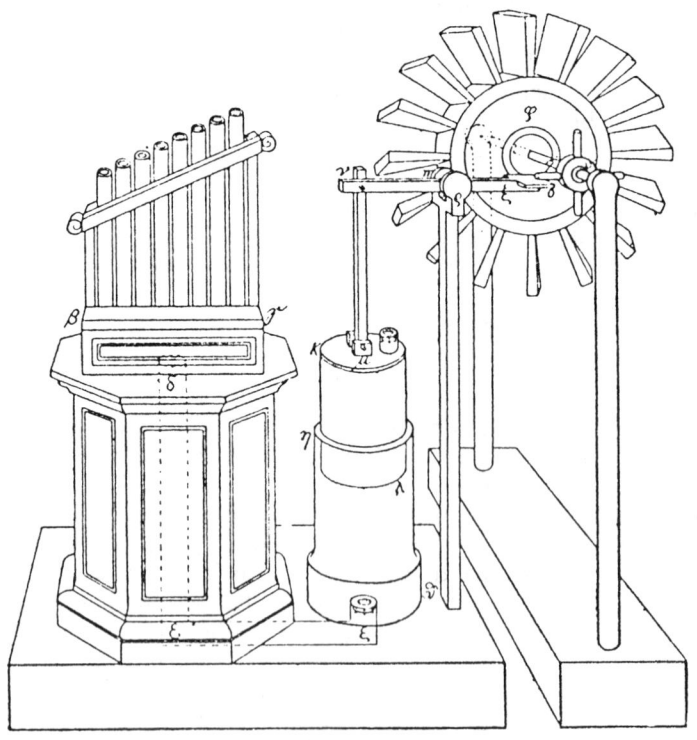

Figure 2 Schmidt's reconstruction of Hero's organ (note 10).

side, each piston being moved up and down by a lever operated by an organ blower. All Hero has done here is to invert the single cylinder and replace the operator by the windmill.

Whether Hero's adaptation would work is quite another matter. It would depend on the relative size of the windmill and on the number of trips being small enough to give time for the piston to descend by gravity before the next trip came into action. He does not specify the number beyond mentioning pegs in the plural, and the manuscript diagrams show anything from four, as here, to sixteen. Two would seem the best compromise. But overall the drive to the pump does not inspire confidence, and the windmill was probably only a paper exercise, never developed in practice. Whatever its ultimate influence, which will call for discussion later, the only direct hint that Hero's design had any effect on others is an echo in *The Book of Ingenious Devices* of the Banu Musa, those engineers of ninth-century Baghdad who were cast very much in Hero's mould and owed a considerable debt to him and his Alexandrian predecessors. They have an alternating fountain worked by wind power: 'On this stanchion we erect splits [Hill interprets these as wheels split into separate vanes] which turn stanchion NM if the wind blows, like those which people are accustomed to install in *anburia*'—this last word being obviously derived from Hero's *anemouria*.[11] Later on we will consider who these people were.

Our main concern is what inspired Hero to produce his windmill design. The crux lies in the sentence 'let the disc ϕ have blades like the so-called *anemouria*'. The word I translate as blade means anything flat, and especially the blade of an oar. Is it the individual blades or the whole circular construction that resemble the *anemouria*? Scholars have been beguiled by the later but (as we shall see) misleading appearance of *anemourion* in the works of Eustathius, where it means a wind vane or weathercock. Since the whole wheel is quite unlike a wind vane, they have assumed that *anemourion* means an individual blade, slender though the resemblance is.[12] In fact the phrase should be read, 'let the disc have blades as do the so-called *anemouria*'.

Writers of literary Greek clung to a formal language which, as the years passed, became more and more out of tune with the spoken tongue, and when they had to use a colloquial term they tended to prefix it with an apologetic 'so-called'.[13] The *anemourion* therefore existed before Hero wrote. It was familiar to his readers, and it resembled the wind-wheel. I suggest that it was a rotary fan, hand-powered, with angled blades; which is no more than a windmill in reverse. Hero stresses that the frame carrying his windmill was separate from the organ and could be turned to face the wind. A rotary fan mounted on a similar frame would be virtually identical; indeed, if stood in the wind, it would act as a windmill. On machines like winches where we would apply the crank, the Greeks and Romans used four projecting handspikes. A rotary fan could easily be turned in this way: if large by hand, if small by finger. The four trips of the diagram could be in direct imitation of such handspikes, but driving rather than driven. The whole thing would be a mechanization of the traditional fan waved by hand, which was age-old and very popular in Egypt, especially with women.

The word *anemourion* is based on *anemos*, wind. *Anemoura* in vernacular

Greek meant a gale.[14] A cognate verb *anemourizo*, to run like the wind, is still found in the dialect of Khios.[15] The suffix *-ion* is a diminutive. *Anemourion* therefore denotes a little gale, and is first recorded as a place name in the 330s BC for the town and cape (still called Anemuri) on the southernmost point of Asia Minor opposite Cyprus, named after its windiness.[16] And for a fan, 'little gale' is no bad soubriquet.

Only once more is *anemourion* found as a fan. A certain Theophanes, who was probably on the staff of the Roman governor of Egypt, some time between 317 and 323 made a journey from Upper Egypt to Antioch in Syria. A long papyrus contains various lists and accounts connected with this journey. At one point it details purchases made en route, mainly of food and wine, but including souvenirs and articles of toilet. What interests us is a list of items, bought perhaps at Heliopolis near the present Cairo, which includes a flask of oil for the bath, 5 lb of cheese, dried figs, combs, thread, sandals, pumice-stone and sponge. Some words are damaged, some abbreviated, some hard to translate. After the oil comes *anemour. bibrad*.[17] The first word, since *anemoura*, a storm, will not do, must be *anemourion*. Theophanes would hardly buy a wind vane on his travels. But, as the editors see, a fan would not be out of place. *Bibrad* . . . is not Greek, but the document contains a number of loan-words from Latin. The editors suggest a connection with *vibrare*, to oscillate. The Latin *v*, absent from the Greek alphabet, was commonly transliterated as *b*; *t* and *d* were to some degree interchangeable, *Bibrad* . . ., then, could be the Latin *vibrat* . . ., perhaps given a Greek termination such as *bibradikon*. The result is not a rotary fan but an oscillating one, a punkah; which is indeed how Liddell and Scott (the standard Greek lexicon) interpret it in their supplement. Even if *bibrad* . . . is something else beyond our understanding, the *anemourion* is still likely to be a fan.

Other evidence for mechanical fans is sparse but real. One possible hint, under a different name, is found in the works of the great physician Galen, writing in the mid-second century AD. For nursing patients who need to be kept in cold conditions, he says, prepare a cool room, preferably underground and facing north; damp the floor and 'blow in an artificial draught from some kind of *euripos*'.[18] *Euripos*, which normally means a strait or narrow sea with vicious currents, must here be a fan, and, by implication, something more forceful and regular than the ordinary type flapped by hand. Indeed the word means something like 'good blast', being derived from *ripis*, a fan. But how this one worked we cannot say.

We have two other names. Around the middle of the sixth century Olympiodorus, a neo-Platonic philosopher of Alexandria, wrote a commentary on Aristotle's treatise on meteorology. It includes this passage:[19]

> This whirlwind the poet [Homer] calls *thyella* (storm); Aristotle calls it a *typhon*, like *typon* (smiting), because it smites mightily and smashes solid bodies; sailors call it a *siphon* (pump) because like a pump it sucks up seawater; the Alexandrians in the local dialect call it an *anemosourin* from its likeness to ladies' *kyklanema*, which the locals call *anemosourin*; doctors call it a *borborygmos* because it resembles that condition in the coils of the intestines.

This gives us two new words, neither of them found anywhere else in ancient or medieval Greek. Both must in the context mean fans. Liddell and Scott (supplement) give for *anemosourin* 'ladies' fan . . . whirlwind . . . (Alexandrian word; perhaps corrupt for *anemosyrin* (from *syro*) or *anemourion*).' *Syro* means to push, and 'wind-pusher' would be a good name, but neither derivation is valid because in modern Greek *anemosouri* means a violent storm or cyclone: Olympiodorus's precise word has remained in daily use although no written record of it has survived from the intervening centuries. What the etymology really was we do not know. *Kyklanemos* is built up of *kyklos*, circle or wheel, and *anemos*. It cannot, unfortunately, denote a wind-wheel, but means wheel-wind or, as we would say, whirl-wind. Once again, what a good description of a rotary fan. In the sixth century, then, Alexandrian ladies cooled themselves with fans called 'whirlwinds' and nicknamed something like 'cyclones'.

Here this particular trail goes cold. Less than a century after Olympiodorus wrote, the Arabs conquered Egypt and Greek culture there died out. Whether the rotary fan was exported to lands that remained Byzantine we cannot say—there is no evidence for it—although one could well imagine it cooling emperors enclosed in their palaces in the sultry summer heat of Constantinople, just like their counterparts in China. Rotary fans there, used both for winnowing and air-conditioning, go back to perhaps the first century BC, but they were always tangential-flow, not axial.[20] The point to underline is that the classical world—or at least the Egyptian part of it— invented the axial-flow rotary fan, and Hero had the mother wit to see that by reversing its action it would serve as a windmill. There is, however, no sign that he or anyone else in the ancient world put it to useful work.

The confusion about wind vanes needs to be cleared up. Not uncommon in the ancient world (though we do not know their name), they became proverbial for people who kept changing their minds. Dio Chrysostom in the first century AD speaks of fabric streamers.[21] Of three monumental versions, one adorned a magnificent circular aviary in Italy, another was on the famous Tower of the Winds in Athens (both of the mid-first century BC)[22] and the third was the Anemodourion (literally wind-spear) in Constantinople.[23] Built somewhere between the fourth and eighth centuries, this was an ornate four-sided structure, visible from all over the city and crowned with a pivoting female statue.[24] Modern Greek has two words for wind vane: *anemodeiktes* (wind-indicator) and, like the Byzantine monument, *anemodoura*, which can also denote a whirlwind, a fickle person, and a windswept place (whence Anemodouri as a place-name).

The wind vane also crops up in the writings of Eustathius, archbishop of Thessalonica *c*.1179–1195/6. When attacked for his style of leadership, he replied accusing his accusers of constantly changing their minds. He compares them to 'the ever-turning *anemourion* . . . swinging and swinging back at every breeze'.[25] In the context he can only mean a wind vane. The reason is that at this time *anemourion* was coming into popular parlance as a synonym for *anemodourion*. A dictionary of the twelfth century (which we know that Eustathius used) states that the word should be *anemourion* and not *anemodourion*.[26] But the usage by no means became universal.

There was yet another word for the wind vane, *aneme*. In late medieval

Greek it ended up denoting the windle, the reel on which skeins of wool were placed for feeding on to bobbins for weaving.[27] But it first appears in another sense, in the fifth- or sixth-century *Apophthegmata Patrum*, a collection of anecdotes of the desert fathers of Egypt. Makarios the Great, it says, met Satan coming back from tempting monks in their cells. 'You didn't find any friends there?' asked the saint. 'Yes,' replied Satan, 'I've a monk there who's a friend. At least he obeys me, and when he sees me he swings round like an *aneme*'.[28] Not, surely, like a windle but (with its usual connotation of fickle-mindedness) like a weathercock. When later *aneme* came to mean a windle, it also took with it the name *anemouri*, for on Paxos today the two words are synonymous.[29]

THE PERSIAN WINDMILL

The first true windmills were in Persia, and since they have been well documented[30] only a few aspects demand consideration here. Their home was the provinces of Khorasan and Sistan, a north–south corridor which follows roughly the modern Afghan–Iranian border, though in 1219 and 1414 they were found as far north as Samarqand. There are reports of scattered specimens both west of the corridor and east as far as Kabul, Ghazni and the present Pakistani province of Dera Ismail Khan in the Indus valley, which in Harverson's view, if they existed at all, were experiments that failed. In the corridor the land is largely flat, with few of the horizontal water-wheels which are otherwise normal to the region; and from the north blows the strong 120-day wind which turns these mills. The medieval sources are Arabic or Persian, plus a couple of Chinese ones. We first hear directly of the mills about 947 both from al-Masudi and al-Istakhri, who give few details. Several more geographers notice them, the most useful account being by al-Dimashqi writing about 1271. He describes a square building with a funnelled vertical slot in each side to lead the wind in to the sails. These, six to twelve in number, were mounted on a vertical shaft and covered with cloth. On the floor above were the stones, the runner being driven directly by the windshaft. By 1414 when Chhen Chheng saw these mills and gave a similar description, the position had been reversed, with the stones below and the sails above, as seems to have been the case ever since; but he still speaks of wind-slots on all four sides. A few years earlier, however, Guzuli had remarked that the mills were directed only towards the north wind, which too became standard practice (Figure 3). Possibly in the corridor proper they were always unidirectional, the four-way ones being limited to its fringes. These mills, latterly at least, never had a roof over the sails.

Grinding was not their only function. Some at Samarqand in 1219 pounded rice by means of vertical stamps, others on the plateau raised water for irrigation (al-Masudi *c.* 947, *Tārikh-e Sistān*, *c.*1062) by means of 'lifting wheels' (Guzuli *c.*1400). Both processes, on the face of it, demand a right-angled gear drive off the vertical shaft. Harverson supposes that it drove a chain of pots.[31] Certainly the horizontal irrigation mill in China, borrowed from Persia in perhaps the thirteenth century[32] (although its

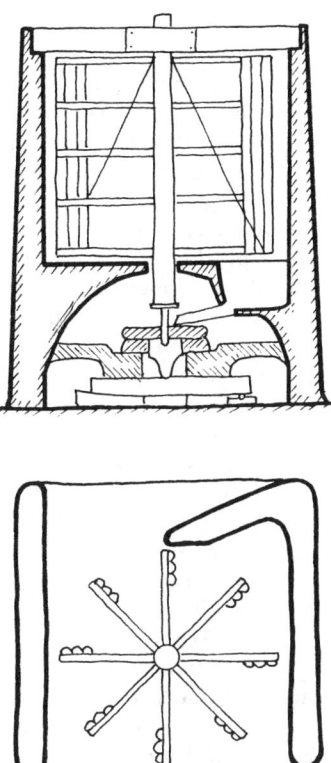

Figure 3 Persian windmill (after H.E. Wulff, *The Traditional Crafts of Persia* (Cambridge, Mass., 1966).

luffing sails were of very different design), was geared to a chain-pump. Yet a nineteenth-century traveller gives us pause. H.W. Bellew, who was in Sistan in 1872, wrote, 'In some parts of the country these mills are adapted to work horizontally for the raising of water, but we did not see any of these.'[33] None has been reported since. While he might imply a right-angled gear, might he not equally mean that the axle lay in the horizontal plane where it could carry a chain of pots—or work rice stamps—without gearing? This arrangement seems to be visible in a photograph of 'A windmill in Seistan' published without comment by G.P. Tate, who surveyed the country for many years between 1886 and 1905. The mill is unlike any normal one, for the building is roofed and the only sail visible has vertical, not horizontal, spokes.[34] A further practical point is relevant. The original horizontal corn mill had the stones above the sails. In a horizontal water-lifting mill, since it would be exceedingly cumbrous to take the drive from above the sails, surely it would be taken from below. Would not builders of corn mills recognize the benefit of placing the sails above the stones and follow suit? Yet as late as 1271 they still had not done so,

which suggests that the irrigation mills had not a vertical axle but a horizontal one.[35] Such a mill would resemble the nineteenth-century Jumbo
mill or 'go-devil' much used for pumping on the Great Plains of the USA,
which, with one exception, that we will meet shortly, was the only known
application of this principle.[36] Though strictly a vertical windmill, it
differs from all other vertical mills in utilizing a tangential, not an axial,
flow of air.

Is this a clue to the origin of the Persian mills? Needham and others
suggest that they were a by-product of the Tibetan and Mongol prayer
wheel, which looked much like a modern anemometer.[37] Possibly so. Yet
it is also observed that the horizontal windmill, especially in the inverted
form described by al-Dimashqi, is remarkably reminiscent of the horizontal
watermill, still a common sight in Persia and Afghanistan. The Jumbo mill
(let us call it that for want of a better name) is almost as reminiscent of the
vertical water-wheel. The Chinese, who were slow off the mark in adopting
gears, very logically tended to apply the horizontal water-wheel to processes
like grinding where a vertical drive was needed, and the vertical water-
wheel to those like pounding with trip hammers where a horizontal drive
was easier.[38] Can it be that Persian windmills followed the same philosophy, and derived from the two types of water-wheel? Yet the evidence
for the Jumbo mill in Persia remains too slender for the point to be pressed.

In the mid-tenth century, the Arab geographers make clear, the Persian
windmill was unique in Muslim lands. It was also, surely, unique in the
world: it had not yet gone to China; it is hardly conceivable (except as the
prayer wheel) among the nomads of Central Asia; there is no hint of it in
India, and none in the Byzantine Empire. Was it then wider spread at an
earlier date? This brings us to the thorny problem of Abu Lulua.

In 644 the second caliph, Umar, was murdered in Medina by a Persian
slave nicknamed Abu Lulua. The early history of Islam was not chronicled
at the time, but only began to be compiled from oral tradition in the ninth
century. The versions that we have of this event therefore vary. By common
consent, Abu Lulua was a carpenter and smith, he belonged to al-Mughira
the governor of Kufa, and he was allowed to earn his own living on payment
to his master of a given sum. When he complained to the caliph that it was
too much, Umar replied that, with his skills, it was reasonable. He added,

> 'I am told that you claim to be able to build a mill worked by the wind.'
> 'I do', replied Abu Lulua. 'Make me such a mill', said the caliph. The
> slave replied, 'If I live, I will make you a mill that the whole world
> will talk of, from east to west!' Then he left. Umar said, 'That slave
> has just threatened me with death'.[39]

This is the version of that conscientious historian al-Tabari, writing soon
after 900. Al-Masudi's of *c.* 947 is very similar.[40] Al-Maqdisī talks simply
of a mill,[41] yet another tradition of hand mills.[42]

Very different things have been made of this story. Some take it as
evidence, Abu Lulua being a Persian, that the windmills of Sistan existed
in 644. So perhaps they did, but the tale is hardly relevant. Horwitz argued
the opposite, that windmills did not exist because the caliph, in asking

for one, was demanding the impossible, and Abu Lulua killed him in desperation.[43] This does not square with any version of the episode, for Abu Lulua always admits to being a mill-builder. Lynn White, with more reason, feels that after three centuries of oral tradition the story cannot safely be used to prove the existence of windmills.[44] The versions which say where Abu Lulua came from are unanimous that he was a native of Nihawand in Luristan, where in 642 the Arabs had finally routed the Sassanian Persian army and seized the town. One tradition indeed states that Abu Lulua was captured there.[45] This makes every sense, since his subsequent master al-Mughira was present at the battle[46] and could well have acquired him when the spoils were distributed.

Yet another version is still more circumstantial about Abu Lulua's career. Ibn al-Athir, though late in date (1160–1233), is ranked among the best of Arabic historians and had access to sources now lost. Before recounting the windmill story much as in al-Tabari, he says that Abu Lulua hailed from Nihawand but in his youth had been captured by the Byzantines and taken back by the Muslims.[47] If this is true, his first capture must have been during Heraclius's Persian wars of 622–8, and his second in the Arab invasion of Syria from 633 or of Egypt from 640. The implication that he acquired the skill of windmill building while in Byzantine hands is not likely: there is no sign of Byzantine windmills at this date.

It seems best, then, to accept that Abu Lulua lived at Nihawand and was enslaved there in 642. There is nothing to link him to Sistan, which was not invaded by the Arabs until 650–1.[48] Nihawand is in western Iran, a good 1200 km from the windmill corridor. It lies in the well-watered Zagros mountains, now and no doubt then good watermill country. What were windmills doing there? There are only two options open. Either we accept the windmill in the Abu Lulua story and accept their presence in western Persia, for which there is no other evidence, or, preferably, we reject them as a later interpolation into the traditions by someone who had knowledge of Sistan, which means that windmills existed there before al-Tabari's death in 923.

There is, however, one earlier pointer. The three Banu Musa brothers, scientists and engineers, were born somewhere around 800 in Khorasan where their father Musa was successively highwayman, noted astronomer, and close companion of al-Mamun who ruled eastern Persia from Marv from 809 until he became caliph in 813.[49] Although they (or rather the middle brother Ahmad) wrote the *Book of Ingenious Devices* in Baghdad, their origin in Khorasan lends further interest to their alternating fountain which has already been mentioned. Its 'vaned wheel' drove a vertical shaft, and although Hill sees it as a vertical wheel driving the shaft through a worm,[50] no gears are mentioned and horizontal sails are much more likely. When they say 'on this stanchion we erect splits . . . like those which people are accustomed to install in *anburia*' they are surely applying Hero's term to the existing windmills of their native land. If so, it pushes the Persian mill back to at least 800.

In default of any other indications, the windmill was no doubt an indigenous development there. But the technological history of this part of the world for six centuries before 800 is a closed book. From about 130 BC (the precise dates are much disputed) until the Arab conquests of the seventh century the area was ruled variously by the Sakas, Kushans, Sassanian Persians, Huns, and Sassanians again. All but the Persians were nomads, or rather had been before they arrived.[51] But before 130 BC—and indeed for a while after it—a pale light does gleam from the dim pages of the history of Greek Bactria. This was created out of Alexander the Great's conquests and between about 250 and 130 BC, isolated from its Mediterranean cousins, it was an independent state, at its widest embracing the area from Samarqand to Kabul. It survived the loss of Bactria by extending its borders beyond the Hindu Kush into modern Pakistan and north-west India, and it finally succumbed to the Sakas about 30 BC.[52] Loss of independence was not the end of the story, for the Greek population survived and the Greek language continued to be spoken at least in India until something like AD 200, and Bactrians were still trading to Alexandria in the early second century.[53] Indeed the new rulers respected and imitated Greek culture until the irruption of the Ephthalite Huns into Bactria in the fourth century and into India in the fifth.[54]

The Bactrian Greeks were famous artists, craftsmen and engineers both before and after the fall of their empire. Their complex irrigation system made Bactria a fertile land until the Mongols wrecked it.[55] They were renowned as well-diggers.[56] Automata entirely in the Alexandrian tradition were to be found in north-west India in the first century AD.[57] The great pagoda of Kanishka at Peshawar (around AD 100?), at 194 m the tallest building in the world, was built by a Greek.[58] All the recorded windmills of the area, from Samarqand to Hamun-i Helmand, and those reported from Kabul, Ghazni and the Indus valley, fall within the borders of Greek Bactria and, coincidence though it may be, it is tempting to trace their history back to this period. Of hard evidence, however, there is none, and short of some miraculous archaeological find there is not likely to be. The origin of the Persian windmill must remain an open question.

The Persian mill found its way, as we have seen, to China. It reached India, where there was one at Delhi by 1320.[59] And Persian sailors perhaps took it to Java or Sumatra where al-Idrisi records windmills in 1154.[60] Whether and how it reached the Mediterranean is our next concern. Muslim Spain is sometimes pointed to as a likely destination or indeed a likely intermediary between the East and north-western Europe.[61] True, al-Himyari says that windmills were established at Tarragona 'by the men of former times'. But while the date of his book is often given as 1262,[62] the surviving version is two centuries younger.[63] Certainly Gille's remark[64] that Spain had windmills by the tenth century is without foundation. By a similar token, claims have been made that there were windmills in Portugal by the eleventh century when the poet Ibn Mukana, who lived near Lisbon, talks of 'a mill turning with the clouds'.[65] This need mean no more than that he had read the Arab geographers on Sistan. Portuguese horizontal mills are mentioned from the late seventeenth century, and a version under that name that was built in the West Indies in the early eighteenth century does bear some resemblance to the Persian archetype. But some believe these mills in Portugal are mythical; there is nothing to grasp hold of, and no dates.[66]

In the eastern Mediterranean, one positive claim for the horizontal windmill must be rejected. Forbes asserts that from Persia it spread west to become 'a feature of the Egyptian sugar-cane industry and thence travelled to the West Indies when Arabic experts from Egypt were lured there to help and establish the first sugar plantations'.[67] But detailed studies of the sugar industry, drawing on medieval Arabic sources for Egypt and colonial archives for the West Indies, offer no support at all for Forbes' statements;[68] and an observant visitor in 1512 stated baldly that in Egypt 'there are neither watermills nor windmills'.[69]

THE BYZANTINE HORIZONTAL WINDMILL

Horizontal mills are not efficient since only a small proportion of their sail area is presented to the wind at a given moment, in contrast to the vertical type where (except for the shielding effect of the mill body) all the sails are always in the wind. In the case of Persia and its violent 120-day wind, this defect hardly mattered. But in Europe it did. Here horizontal mills were sporadic in location and in date, and only in one small area did they dominate for any length of time. They fall into three categories.[70] There were engineers' designs from the Renaissance onwards, often quite elaborate, many of which probably never left the drawing board. Next, there were simpler workaday versions invented or reinvented by local craftsmen to serve a local need, and in general of relatively recent date. Neither type seems to owe allegiance to the mills of Persia, except perhaps in ultimate inspiration by hearsay at many times removed. None effectively caught on. The third category was found in the Byzantine Empire, the only area where a long-established tradition is discernible, and where a direct debt to Persia is a very real possibility. There are five examples in this group, most of them hard to date but, when taken together, suggestive.

The first are the horizontal mills which worked until recently on the island of Karpathos between Crete and Rhodes; no other surviving examples have been reported from the Aegean. They were unidirectional and simple. The house was square; the shallow lower floor contained four horizontal sails, with an opening in the front and back walls to direct the wind through; on the upper floor were the stones (Figure 4). Being the only known horizontal mills which in this respect resembled the Persian mill described by al-Dimashqi they may very well represent an early version little changed from the original model. Their latterday name, *taralis*, means the churn-staff for churning milk, a vertical rod with four blades projecting at right angles which is rotated between the hands: a good analogy for the design. But the history of the *taralis* mill beyond living memory is not on record.[71]

Also unique, apparently, to Karpathos and also called the *taralis* was the Jumbo mill. It was necessarily unidirectional. Just as the horizontal mill

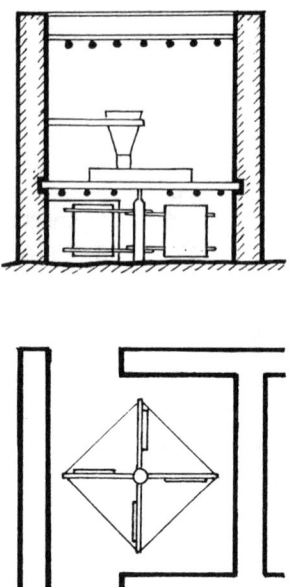

Figure 4 Karpathos *Taralis* (after Limona-Trebela) (note 71).

had two compartments one on top of the other, this had two side by side. The sail compartment, roofed over, was not unlike a vertical water-wheel pit, for nearly half the sails were shrouded by being below ground level. Four sails turned a horizontal axle which passed through to the milling compartment where it drove the stone by a twisted belt drive.[72] To link this mill with the Jumbo type considered in Persia is tempting but probably illusory. It is said that the Karpathos mill was not efficient; to use the Jumbo rather than the horizontal version for grinding seems perverse since the direction of the drive has to be changed; and the belt drive does not smack of any great antiquity. Harverson suggests that it was of quite recent build, inspired by a Greek emigrant returning home from America who had seen the genuine Jumbos there.[73] This explanation rings true.

   Next, the Crimea, where the direct evidence is not very old. In 1793–4 Pallas observed horizontal mills both in the Crimea and at New Nakhichevan near the mouth of the Don. The sails, mounted atop a circular structure, were shielded from the wind on one side by a movable segmental screen. The mills at Nakhichevan were recent, built by Armenians who had emigrated from the Ottoman-ruled Crimea to Christian Russia in 1778.[74] At the same time there was a mass exodus of Greeks to the north shore of the Sea of Azov. With the departure of both these peoples, the Crimean horizontal mill languished. The last example, long derelict, at Eupatoria (Yevpatoriya) was recorded in 1854 with some detail, notably that the sails were fitted with pivoting shutters; altogether a sophisticated machine.[75] In Pallas's words, the mobile screen (and one might guess the shutters too) were 'an invention the merit of which is claimed by the Armenians'. But if the Armenians were improvers, were they originators? They had settled in the Crimea from the thirteenth century, but no hint has yet been found of windmills in Armenia proper, while the Crimea had been a prosperous and strategic Byzantine possession.[76] In something like the 1120s—precise dates are not recorded—much of it was overrun by nomadic Cumans, and Byzantium clung on only to the towns of the mountainous south-east coast, which is watermill country. The steppes of the north and west, the windmill land, remained in Cuman control for a century until the whole Crimea was conquered by the Tartars, who permitted the Genoese and Venetians to establish trading stations in the east of the peninsula.

It is tempting to ascribe the Crimean windmills—and many others that we will meet—to the Italians. But all such links founder on the simple but surprising fact that Italy has never, at any date, indulged seriously in windmills. Although much of the country is obvious watermill territory and the Po valley went in for boat mills, one might expect windmills in other low-lying parts like the Adriatic coast from Ravenna to Trieste. Not so: there is a very dubious reference to one (at Siena) in 1237, two during the next century, and correspondingly few thereafter. While Dante's allusion to a windmill show that educated Italians of the early fourteenth century knew about such things, one authorized at Venice in 1332 sounds like a novelty there.[77] Moreover, the Black Sea had long been a Greek lake. The Greeks were masters of its waters and of much of its coastline, the population of which, though an ethnic jumble, was dominated by Greeks. Throughout the countless invasions which swept around it, these Greeks provided the only thread of continuity. The permanent and sole civilizing influence, they tried to teach an agricultural way of life to the invading nomads. The Venetians and Genoese, when they finally settled in the Crimea, came merely as traders in search of profit and stayed for a relatively short time, in small groups closeted in a few mercantile centres. The Greeks, in contrast, though dwindling in number in recent centuries, have lived and farmed all the way from the Danube to the Kuban for two and a half millennia.[78] It seems much more probable, then, that the essentially rural technology of the windmill was introduced to the Crimea by the Greeks than by the Italians; and quite possible that the horizontal windmill, in simple form, was introduced before about the 1120s when direct contact with Byzantium was lost. If this is so, its long survival was no doubt due (as in Persia) to the notoriously strong winds of the Crimean plains.

Next comes a much better record in terms of dating. In 1486 Konrad Grünemberg, on pilgrimage from Switzerland to the Holy Land, stopped off at Kandia (Iraklion) in Crete. There he saw

> a very strange mill. It lies in front of the town near the sea. It has eight sides and each side has a door. Wherever the wind comes from, a door or two is opened in the same direction. For there is a leaf [i.e. a sail] on the other [side] made of cloth. The wind drives it, and so that the wind can escape, another door is opened, so that we saw the whole thing turning efficiently and grinding. This is a drawing of it (Figure 5), and where the millstone turns is a cavity dug in the rock below.[79]

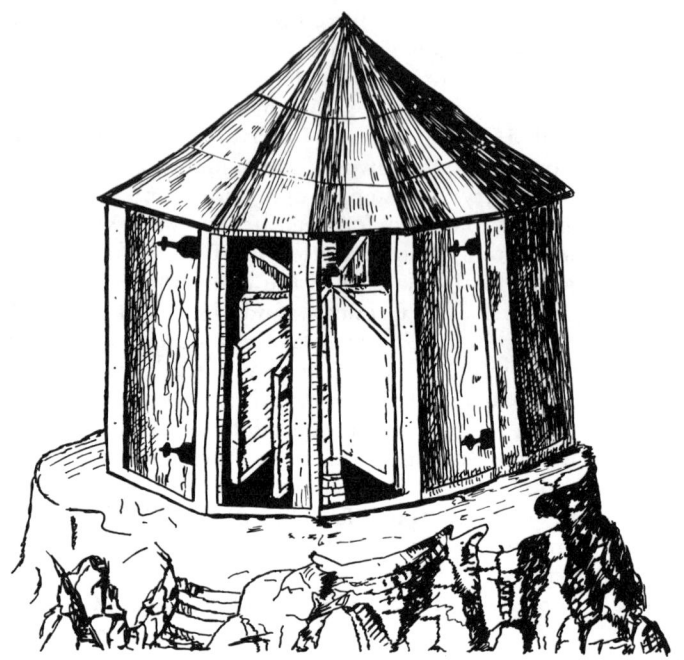

Figure 5 Kandia horizontal mill 1486 (after Grünemberg) (note 81).

A Spaniard, Pero Tafur, had likewise landed at Kandia in 1436 and remarked the 'many windmills' by the harbour.[80] Whether they were horizontal he does not say; but probably they were, because no vertical mills are to be seen in contemporary views of Kandia.[81] Once again, how long had they been there?

The fourth instance, though not in Byzantine lands, may yet be derived from them. At Borgio Verezzi near Savona in Liguria is the cylindrical masonry housing of a horizontal mill, with eight angled and tapered wind openings on the upper floor (Figure 6).[82] It falls into the same sort of category as the Kandia mill. The date, though obviously old, is unknown; but Savona is near Genoa, and the Genoese had very extensive commercial interests in the Aegean after 1155, even more so after 1204, and in the Crimea from 1266. In the absence of other early evidence for the horizontal windmill in Italy—and even for significant numbers of vertical ones—it seems much more likely that the Genoese copied what they saw in the Aegean or Black Sea than that they inspired the mills there.

The origin, then, of the first four examples cannot be dated, although a case can be argued that the Crimean mills were introduced before the 1120s. We are on firmer ground with the last example. This is to be found in a semi-comic French *chanson de geste*, the *Voyage de Charlemagne à Jérusalem et à Constantinople*. It tells of Charlemagne's totally fictitious pilgrimage to Jerusalem in search of holy relics and of his return via Constantinople,

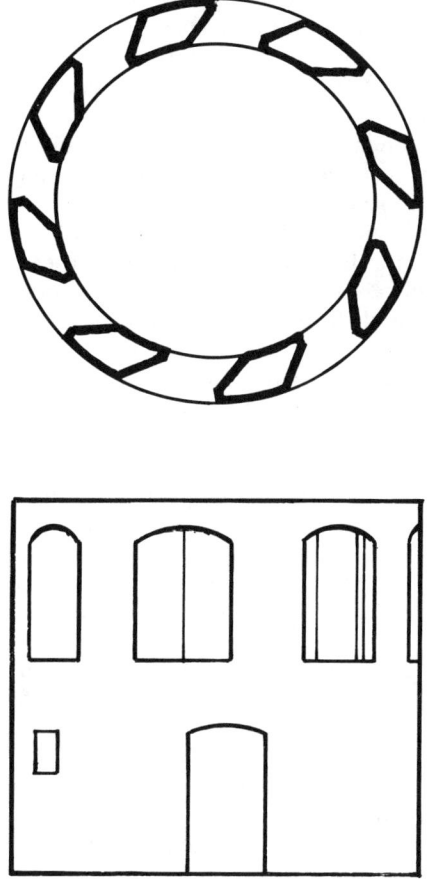

Figure 6 Borgio Verezzi horizontal mill (after Wailes) (note 36).

where he is given the friendship and ultimately the allegiance of King Hugon. The palace where he is entertained is a marvel, described with many details of which only the most relevant can be quoted here.

The palace was vaulted with a dome on top;
It was built with great care and put up with art;
Its central column was inlaid with silver,
One hundred pillars of solid marble stand there,
Each one inlaid with fine gold in front.
Two brass and metal children flank it,
Each holding in his mouth a horn of white ivory.
If there blow from the sea nor'westerlies, northerlies or other winds
Which strike the palace on its western side,
They make it turn as swift and continuous

As the wheel of a vehicle going downhill.
These horns then resound and bellow and boom
Like drums or thunder or great hanging bells;
Then [the statues] face each other and laugh
So that you would think they were alive . . .
[Charlemagne] hears a wind coming from the ports,
It comes raging against the palace, strikes it on one side,
Sets it in motion gently and smoothly,
Makes it rotate like the shaft (*arbre*) of a mill.
The statues blow their horns, laugh at each other . . .
[Charlemagne] cannot stand on his feet, and sits down on the marble.
The French are all thrown down, they cannot stand up . . . [83]

Such a magic roundabout of a palace never of course existed, at Constantinople or anywhere else. But in principle it is a horizontal windmill, and apparently a unidirectional one at that. Such a concept would hardly be dreamed up out of the blue, and there was nothing like it in French literary or mechanical tradition on which the poet could draw. It is clear, in fact, that his inspiration was Byzantine. Many features of the palace reflect Byzantine themes: its decoration, the great carbuncle that lit it, the automata blowing horns, the furniture all suggest the author drew on some lost Byzantine romance. Even the windmill finds some parallel in a huge suspended pearl, rotated horizontally by the wind, in Meliteniotes' *Sophrosyne*, a later poem but in the same romantic mould.[84] The mention of 'the shaft of the mill' is probably a simile drawn from a water or horse mill and inserted by the French poet to elucidate the principle to an audience unfamiliar with horizontal windmills. The whole fancy, indeed, was perhaps originally inspired by the name of the imperial palace at Constantinople, the Magnaura. Actually a corruption of *magna aula*, great hall, it was popularly supposed by the Byzantines to mean 'great breeze'.[85]

Although the *Voyage* has been variously dated, opinion now firmly favours the twelfth century. Its description of Jerusalem accords with the period 1120–1150 and no later.[86] In 1146, when the Second Crusade was in preparation, the emperor Manuel Komnenos demanded of Louis VII, should he come crusading, an oath of allegiance. This haughtiness was not well received in France. In the *Voyage* the tables are turned; Hugon turns out to be not so superior after all, and ends by swearing allegiance to the French king. The *Voyage*, I suggest, was composed at this time to cock a political snook at the Byzantines, and for local colour drew on a Byzantine romance.

It seems, then, that by the mid-twelfth century Constantinople knew of the horizontal windmill, and well enough to incorporate the principle in a popular romance. I suggest that the horizontal mills of Karpathos, Kandia, Savona (via Genoa) and the Crimea belonged to the same old-established Byzantine tradition. Short of independent invention, its ultimate source can only have been Persia. But until 1204 the Aegean was, with one brief exception, wholly Byzantine, and between it and Sistan lay the huge and now fragmented mass of Islam. While transmission of the idea by indirect hearsay is possible, such is the similarity of the Karpathos mills to their early Persian counterparts that a more positive contact should be sought. The clue is to be found in that short period when Byzantium lost part of the Aegean coast. The suggestion that follows, though necessarily speculation, does fit the apparent facts.

In 1037 a horde of Turcoman nomads led by the Seljuq Turks moved south from the steppes of Central Asia to seize Khorasan and Sistan. They then swept westwards and by 1055 were at the gates of Baghdad, where the tottering Abbasid caliphate welcomed them, and became their puppet. Within a decade the Seljuqs had added Armenia to their rule and in 1071 routed the Byzantines at Manzikert. Byzantine Asia Minor, as ripe for plucking as the caliphate, lay open. By 1075 they had overrun it, all but a little, reached the Aegean coast, and set up the capital of their new Sultanate of Rum at Nicaea (Iznik) barely 100 km from Constantinople. For the first and last time since the days of Alexander the vast swathe of territory from Aegean to Hindu Kush was under a single sway. But this unity was short-lived. In 1097 the Crusaders drove them from Nicaea and the western third of Anatolia, which was returned to Byzantine rule. Although it remained a nominal entity until 1157, the Seljuq empire rapidly fell apart. An independent Sultanate of Rum nonetheless survived and flourished until the Mongol invasions of a century later and the rise of the Ottoman Turks after that.

The impact of the initial Turcoman wave was no doubt unpleasant for those at the receiving end, but the Muslim Seljuq rulers were widely welcomed, for, compared with the grasping Byzantine absentee landlords, their regime was enlightened. They protected the existing Christian farmers and moved in peasantry from other lands they had overrun. In 1080, in the wake of the conquest, a great immigration took place of peasants, tradesmen and artisans, some of them at least from Persia. By the mid-twelfth century Asia Minor was prospering mightily in agriculture, industry and commerce. The Seljuqs encouraged arts, science and literature; they built roads, bridges and caravanserais; and it was claimed that for the first time in centuries not only convoys but solitary travellers could journey in safety from the Mediterranean to beyond the Oxus.[87]

Western Anatolia was under Seljuq rule for too short a time to benefit much from this boom, but I suggest that in this brief period the windmill was brought west from Khorasan and planted on the shores of the Aegean by the migrants whom the Sultan encouraged to follow in the train of the army. Hard evidence there is none. All records of the Seljuqs are thin, and of their agriculture and technology they are virtually non-existent. In any case Asia Minor is essentially watermill country where only occasional windmills can be expected. This window of opportunity opened in 1075, and with the fragmentation of the empire it effectively closed well before the mid-twelfth century. Indeed, since the earliest specifically recorded windmills in Asia Minor, as we shall shortly see, were in territory lost to the Seljuqs after 1097, one might argue that the window closed at that date. There has been no other time since the third century BC when peaceable people with peaceable ideas could travel so freely from Khorasan to the

Aegean; and if the windmill was established there by the mid-twelfth century, its arrival in 1075–97 is not unreasonable. One other consideration, of function rather than of design, is suggestive. The known Anatolian windmills of the thirteenth century lifted water. Mills were doing this in Sistan three centuries earlier, but did not do it in western Europe until two centuries later, which makes an eastern derivation more probable than a western one.

THE BYZANTINE VERTICAL WINDMILL: GREEK SOURCES

On the mountainous mainlands of the Byzantine Empire as it was then— Greece, Asia Minor, the Balkans and southern Italy—the watermill has long been predominant despite the seasonal nature of the lesser streams. It was on the Aegean coast of Asia Minor and on the Aegean and Ionian islands, where streams are generally meagre and winds stronger, that windmills found their role. The distribution map of relatively recent windmills in Greece and Turkey (Figure 16, p. 171) is doubtless incomplete, but it is accurate in locating their centre of gravity in the southern Aegean. What of earlier days? Windmills are never once mentioned in Byzantine literature in the ordinary sense, which tends to be orientated towards Constantinople and to show little interest in the provinces, or indeed in technical matters of any sort. A parallel instance is mining which, despite its obvious strategic and economic importance, is largely ignored by Byzantine literature.[88] For windmills, until about 1400 the only documentary evidence to go on consists of ecclesiastical deeds. Compared with that of north-western Europe, the corpus of Byzantine archives is very small—there is nothing comparable, for example, to manorial records— and the vast majority of those we do have are monastic. On the rare occasions that they mention windmills they say nothing about their design. Although most if not all were probably vertical, some might have been horizontal. While we cannot be certain until illustrations begin to appear, it seems best to include them all in this section.

The first windmills I have identified in Byzantine lands were in Asia Minor. They belonged to Lembiotissa monastery near Smyrna (Izmir), which in 1228 was refounded and richly endowed by John III Vatatzes of the Empire of Nicaea, that fragment of the Byzantine realm that survived the Crusaders' capture of Constantinople. Among the possessions listed in his charter to the monastery is a site at the mouth of the River Hermon (now Gediz) about 30 km west of Smyrna. Large, but of very variable flow and much inclined to change its course, the Hermon ended in an alluvial plain sprinkled with lagoons and intersected by regulatory canals. By 1228 the monks had already made one lagoon into a fish-pond called Gyros, which gave rise to a number of legal wrangles over water rights with the owners of nearby fish-ponds and of the rights to the river itself.

The archives recording these disputes make frequent mention of the monks' 'windmills with their canals', which must have been for lifting water, whether to supply the fish-pond in dry seasons or to prevent flooding in wet ones. They are not called by the standard later name of *anemomylones*,

Figure 7 Mills on Gulf of Smyrna 1685–7 (after Lemerle) (note 91).

windmills, but (in the singular) *he ex anemou* or (plural) *ta ex anemou*. In each case *mechane* or *mechanemata*, machine(s), is understood—'the machines powered by wind'—but no more details are given. This circumlocution is perhaps because the word *mylon* was still reserved for a grinding mill. The wind-pumps are first mentioned in a document of April 1234, when they had been in existence for a few years, and there are intermittent references until 1284.[89] For grinding, the monastery owned quite a number of watermills in the Smyrna region. Lembiotissa was destroyed in 1307 by the Ottomans.

The winds here blow from every quarter in roughly equal measure,[90] and at least in later times these mills were probably of vertical type. We may have an illustration of them or their successors. A French chart of the Gulf of Smyrna made in 1685–7 shows as landmarks on the shore near one of the mouths of the Hermon a row of five objects. Each has a vertical post with six radiating spokes on top, distinguished from 'proper' windmills of which four at Karaburun on the opposite side of the Gulf are marked with plump bodies and six sails (Figure 7).[91] The two symbols are remarkably similar to the conventional signs for wind-pumps and windmills on the 1 : 50,000 Ordnance Survey maps of Britain. So perhaps the wind-pumps of the Hermon plain had spindly wooden supports for the sails, though how they were turned and how they lifted the water we cannot tell.

The next batch of windmills is not so certain because the terminology is less clear-cut. It concerns the small monastery of Koteine somewhere in the diocese of Philadelphia, the present Alaşehir, 100 km east of Smyrna and on a tributary of the Hermon. All that we know of Koteine is contained in a single document, the monastic rule drawn up in 1247 by the abbot Maximos, who over the previous decades had built up the monastery and its property in the area to a surprisingly rich level for so remote a house. The detailed inventory of holdings lists no fewer than twelve mills which fall into two sharply distinct categories. One consists of seven watermills (*hydromyloi*) or watermill works (*hydromylika ergasteria*) on five sites; the name of the stream that drove the mill is normally given, and none had any land attached. The other group consists of fields with mills; none is called a watermill. These comprise:

> Field of the mill which came to me by exchange from Ioannes Charmonites.
> Another very small field below the grange [of Aulax] with the old mill, bought from the same, and the old mill belonging to Kazanes with its marked-out land . . .
> Branas' field with the old mill which came to me by purchase . . .
> Field close below [the grange of] St Constantine with the derelict small mill (*mylarion*) which came to me by purchase.[92]

These lands were irrigated, as the plain of Alaşehir still is today:[93] the inventory lists fields with ponds, fields with tanks, fields with canals and sluice gates, fields that are under-watered or waterless. The fields with mills seem to fall into this same class: in other words the mills lifted water. While bucket wheels or chains of pots turned by animal or water power remain a possibility, these were always called wheels, never (to my knowledge) mills. Small windmills dotted over the plain, much as the modern ones at Lasithi and elsewhere in Crete, seem inherently more likely, even if it means that here, unlike Smyrna, they called a non-grinding device a mill. If this interpretation is right, the interesting point is that in three cases the mills are old mills (*palaiomylones*). *Palaio* sometimes means 'abandoned' or 'ruinous', but hardly here, for one mill is specified as derelict by a different adjective. Were they then of outdated design? They could well date back to the twelfth century: were they survivors of the horizontal type imported from Persia, in contrast to the more recent vertical mill?

Windmills for grinding are first heard of on the notoriously windy[94] north Aegean island of Limnos, where a number of the monasteries on Mount Athos came to acquire considerable property. The largest holdings (and the fullest records) belonged to the Great Lavra, three of whose inventories mention seven windmills. Along with vineyards, fields, water-mills and the like, some had been sold or given to Lavra by individuals, even by peasants, who on occasion had shared them with others so that Lavra acquired only a part interest. Other mills still belonged to dependent peasants. One or two are mentioned merely as landmarks on boundaries. These windmills, with the dates they are recorded and the *Lavra* references,[95] were at or near:

Sphouggaras	1304, 1361	ii.99.39, iii.139.44–5
Chleion Neron	1304	ii.99.54
Kontovraki	1304	ii.99.152
Myrina	1355	iii.136.34
location lost	1355	iii.136.119–20
Areione	1355	iii.136.169
Kotzinos	1361	iii.139.88

Another of the Mount Athos monasteries, Dionysiou, also had a windmill in 1430 at Atziki on Limnos.[96]

In 1323 one-third of the village of Mamitzona in the Parapolia—the region immediately west of Constantinople—was granted to the abbot of the Serbian monastery of Chilandar on Athos, its revenues to be devoted to the Serbian hospital in the capital. Included in the grant were two winter watermills and half a windmill.[97] Another windmill on the mainland is recorded in 1362, when a small estate at Skala, probably on the Strymonic Gulf but not precisely located, was given to the Athos monastery of Kutlumus. *Skala* means a landing-place, and the property included a warehouse and a windmill.[98] In all these cases—Limnos, Mamitzona and Skala—the mills are called *anemomylones*, windmills, and were evidently for grain, but no further details emerge. Another mention of 'the old windmill' on Mount Athos itself in 1302 is unfortunately in a forged charter.[99]

Thus far the monastic sources. The windmills are few; overall perhaps one is recorded for every fifty watermills (or unspecified mills, which are almost certainly watermills). No doubt, taking Byzantine lands in general, watermills were very much in the majority. Nonetheless the monastic archives cannot give a representative picture, for they are heavily biased geographically. The largest proportion covers Macedonia and Thrace. There are a fair number for Sicily and southern Italy, some for the Trebizond area and a few for the west coast of Asia Minor and for Morea. All of these regions are essentially watermill country. For the real home of the Greek windmill in more recent times, the islands, the archives are desperately thin or totally lacking. The sole exception is Limnos where, as we have seen, eight windmills are on record, for in this one case the archives are relatively full. There are a few deeds for Khios, Kos and Leros, and very few indeed, considering its size, for Crete. Small wonder, then, that windmills appear so rarely.

Another important factor in assessing the archives is that of date. The great majority go back no further than the thirteenth century, while those that do survive from an earlier time refer, in the main, to watermill areas. For Lavra's holdings on Limnos, for example, the earliest detailed document concerns its grange of Gomatou in 1284 and includes two watermills. The next, also for Gomatou, is that of 1304 which includes five watermills and three windmills. This does not mean that the newly recorded mills were newly built, but that they had been acquired by the monastery during the previous two decades. How far back they went we have no means of telling. In 1304 the lands belonging to Gomatou totalled about 18 sq km.[100] The area of Limnos is 477 sq km, so that by the crudest of yard-sticks one might reckon at that date about eighty windmills on the whole island. Not all of it being equally cultivable, the number would be less, but no doubt still considerable. Windmills very likely existed here in 1284 (and who knows how much earlier), before Lavra acquired any. The records of the Limnos holdings of other monasteries start even later. By the same token the archives of Lembiotissa near Smyrna only begin with its refoundation in 1228; how much earlier had windpumps been used on the flatlands of the Hermon?

One final example of the problems. For arid islands of the south Aegean such as Mikonos, Thira and Nisiros, latterly served exclusively by windmills, no monastic records survive at all. One such island, however, did have a flourishing monastery whose archives are in part preserved. When the monastery of St John was established on Patmos in 1088 it was reported that there was no surface water whatever, and that the arable land comprised a tiny fraction of the island's area. It was uninhabited.[101] When the monks and the supporting lay community moved in, being unable to grow enough grain on the island they had to rely on imports from Crete and later from their estates on Kos and Limnos.[102] These arrived, as one might expect, not as flour but as grain. Where was it ground? Watermills were impossible on Patmos. Hand mills or animal mills would be feasible, the latter being not unknown within monastic precincts elsewhere.[103] Or there could have been a windmill, such as Patmos has to this day,[104] as soon as the community became large enough to support it. We do not know. The records are silent because the monks owned the whole island and had no boundaries to define; they had no need of inventories such as were compiled elsewhere where monastic properties were jumbled up with others and precise boundaries were a matter of importance. We therefore know more about the estates of Patmos on Kos and Leros than we do about Patmos itself.

These limitations of the sources must always be borne in mind. In western Europe we enjoy a relative plethora of medieval records. Their collective silence on windmills before the 1180s makes it well-nigh certain that such things did not exist here, at least in significant numbers. In contrast, the silence of the much poorer Byzantine records before 1234 means no such thing. Only the rare statement in an external source allows a more categorical judgement. Michael Choniates, for instance, remarks that in the early thirteenth century the island of Kea produced no corn and that wheat bread was unobtainable there;[105] hence we can deduce with some confidence that Kea had neither watermills nor windmills at the time. But for the rest, windmills on the islands—any of the islands—are by no means impossible in the twelfth century or even the very end of the eleventh. Not all the monastic archives having yet been published, new evidence could still emerge.

THE BYZANTINE VERTICAL WINDMILL: OTHER SOURCES

When we turn to Latin and other sources the picture is somewhat extended in scope, though not in date. After the Crusaders took Constantinople in 1204, most of the surviving Byzantine territory was divided among the victors and entered a troubled period. Of the windmill sites mentioned so far, for example, Smyrna and Philadelphia remained Byzantine under the Empire of Nicaea. Limnos was allotted to the Venetian Navigaiosi family until the restored Byzantine Empire recovered it in 1276. Skala went to the Latin Kingdom of Thessalonica until that was recovered in 1246. Mamitzona was part of the Latin Empire of Constantinople until 1261. Later still, as the Byzantine Empire approached its demise, mainlands and islands were again seized by Turks, Venetians, Genoese, or Byzantine despots, in a complex history we cannot hope to follow here. Venetian and Genoese archives contain much material relating to their possessions of various periods. Little has been published, but Italian documents about the Peloponnese in the fourteenth century, though mentioning only watermills, give a taste of how useful these sources could be for the history of windmills.[106] Somewhat later, for instance, a Venetian report of 1518 refers to windmills on Mikonos.[107] Likewise Ottoman records of the territories they conquered, the study of which has a long way to go,[108] may one day shed light on Aegean windmills from the fifteenth if not the fourteenth century. Nonetheless, it must be emphasized that Latin sources for the Aegean suffer the same limitations as the Greek: they start in detail only in the thirteenth century at best, and mostly only in the fourteenth. Before then, commercial or administrative records of the Italian states, and accounts by pilgrims, are skimpy.

Figure 8 Rhodes 1480 (after Caoursin) (note 114).

Until these non-Greek sources are better known, they illuminate only two areas. One is the special case of Rhodes, where Greek sources are virtually silent. Western ones, however, are quite voluble, both because it was a regular stopping-off point for pilgrims to the Holy Land and because it came into the hands of the Knights of St John of Jerusalem, the Hospitallers. After 1204 Rhodes was held by the rebel Byzantine family of Gabalas until 1249 when, after an abortive Genoese attempt to seize it, it was retaken by the Byzantines. The Knights captured it in 1309, clung on to it despite a prolonged siege by the Turks in 1480, and finally succumbed in 1522. The harbour and town of Rhodes were famous for their windmills: twelve to sixteen on the eastern mole, two or three on St Nicholas' mole, three or four inside the town, and several outside it to the north and west.[109] At least fourteen existed on the eastern mole in 1389 when they are mentioned in the archives of the Knights.[110] A persistent tradition recorded by six pilgrims[111] held that (in Breydenbach's words, which are typical) these mills

> were built long ago by the Genoese. For they tried by deceit and treachery to subdue the city itself and the island; but after many of them had been killed they were punished by this penalty, that they should build these towers and mills at their own expense, for the perpetual benefit of the inhabitants and as an eternal memorial to their own iniquitous treachery.

This can only refer to the Genoese attack of 1249, and though confirmation is lacking there is little reason to doubt the date.[112] These mills would thus be built at the instigation of the Byzantine authorities. All the windmills of Rhodes were apparently for grinding except that, as Nicola da Martoni reported in 1394, the gardens and orchards north-west of the city were irrigated by windmills lifting from wells with bucket wheels.[113]

We have too some early drawings of the windmills of Rhodes. In 1420 Buondelmonti—of whom more soon—shows a token few, with four or six sails depending on the version. Caoursin in 1480 shows the full complement, each with six sails (Figure 8).[114] Breydenbach's well-known view of 1483 also shows them all in considerable detail, some with six sails, some with eight (Figure 9).[115] In all drawings the mills are cylindrical tower mills and all have rectangular-frame sails. The multiple sails go back at least to 1396, when Anglure remarked on 'sixteen windmills all in a row and very

Figure 9 Rhodes 1483 (Breydenbach) (note 81).

close to each other, and most have six sails (*volans*)'.[116] Today only a few survive, with jib sails; whether they are the medieval structures there is no easy way of telling.[117]

I know of no illustrations of windmills by Byzantine artists, who, like writers, were not interested in such things. But Rhodes has introduced us to Cristoforo Buondelmonti, a Florentine priest who from about 1406 spent eight years on Rhodes and six touring the Greek islands. In 1420 he completed his *Liber Insularum Archipelagis*, a description of the islands whose main interest lies in the accompanying maps. The work became enormously popular and was much copied. Six of the maps mark windmills: one on Kefallinia, three on Rhodes, two on Lesvos, twelve on Khios, six at Gallipoli (Gelibolu), and one (or sometimes two) at Pera (Galata opposite Constantinople on the Golden Horn).[118] All his mills are drawn in the same way, a cylindrical tower with conical cap and rectangular-framed sails, four or six in number depending on the copy. His accuracy in marking windmills can be checked in two cases. At Rhodes, given the limited space available on the map, he is correct. Likewise at Khios his four mills south of the town and three to the north correspond to the six-sailers, numbering respectively thirteen and eight, on Braun and Hogenberg's print of *c*.1588 (Figure 10).[119] At Gallipoli, moreover, though Buondelmonti shows only six mills, his adjacent legend reads 'These mills are more than fifty' (Figure 11). He seems, in short, to be trustworthy in the siting and appearance of those mills that he includes. Conversely, he was not drawing Ordnance Survey maps which marked every windmill there was: we cannot assume that an island without windmills on a Buondelmonti map was actually innocent of them.

Later in the fifteenth century more illustrations appear. Breydenbach has a view of Modona (Methoni on the south-west tip of the Peloponnese) in 1483 which includes two eight-sailed mills identical to those he depicts at

Figure 10 Khios *c.* 1588 (Braun and Hogenberg) (note 119).

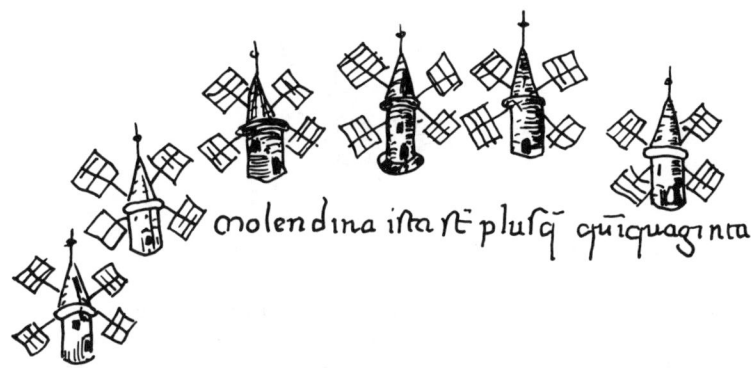

Figure 11 Gallipoli 1420 (after Buondelmonti) (note 118).

Figure 12 Methoni 1486 (after Grünemberg) (note 120).

Rhodes; three years later Grünemberg shows the same, but six-sailed (Figure 12); and two pilgrims confirm their presence in 1480.[120] A panorama of Constantinople from Scutari (Usküdar) published by the Venetian engraver Giovanni Andrea Vavassore, and perhaps based on a drawing done on the spot by Gentile Bellini in 1479, shows a curious pair of windmills in Scutari: each is an ordinary building with pitched roof and multiple (eight?) sails attached to one gable (Figure 13).[121] The same view was borrowed in 1544 by Sebastian Münster, who added two similar mills

Figure 13 Scutari 1479? (after Vavassore) (note 121).

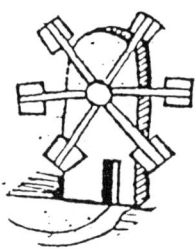

Figure 14 Constantinople 1493 (Schedel) (note 124).

at Scutari and another at Pera,[122] and by Braun and Hogenberg in 1572 who reverted to the original two at Scutari.[123] Finally a view of Constantinople published in the *Nuremberg Chronicle* of 1493 has three windmills in the city itself.[124] They have no cap as such, but a domed top and six rectangular sails (Figure 14).

Topographical prints of this kind and period have to be treated with much caution, for artistic licence was prevalent. Had Vavassore drawn standard western mills one might with reason suspect him, but his odd depiction actually enhances his worth. The mills he shows very probably were fixed vertical ones like those found latterly on Thira, Crete and Karpathos,[125] for at Istanbul 66 per cent of the winds blow from the north and north-east.[126] The presence of mills around the city is confirmed by a report to the Sultan in 1638, which says that Istanbul and its suburbs (including Galata (Pera) and Scutari) had 600 windmills and twenty-eight watermills. The city itself, however, had 985 horse mills, four watermills and by implication no windmills at all.[127] The suburban mills, even if their total is exaggerated, were by this time clearly numerous. We may therefore accept Vavassore with greater confidence, whereas the mills inside the city in the stylized and inaccurate woodcut of 1493 should be dismissed; if they are meant to be fixed mills, they are facing the wrong way. The clustering of windmills seen at Rhodes, Khios, Gallipoli and evidently around Constantinople is paralleled by the report in 1599 of a row of twenty-two six-sailed mills at Heraclea (Marmara Ereğli) on the Sea of Marmara.[128]

Five illustrators, then, of reasonable trustworthiness—Buondelmonti, Caoursin, Breydenbach, Grünemberg and Braun and Hogenberg on Khios—all have tower mills of the type well known in the Aegean in later times, except that in every case the sails have rectangular frames. Of this last feature, more later. Vavassore and his imitators have fixed mills. All were westerners. One might argue that they simply drew the mills and sails to which they were accustomed. Yet the standard mill-image of fifteenth-century Europe was a post mill, and the illustrators' unanimity strongly suggests that they drew what they saw.

Another question concerns the builders of these Aegean mills: in what tradition were they working? Byzantine landowners could build their own mills; so could (and did)[129] Byzantine peasants, for there was no milling soke. Of the windmills recorded up to 1400, pretty certainly all were the work of Byzantines, whether as landowners or tenants (Smyrna, Philadelphia, Limnos, Mamitzona and Skala) or as a municipal or even military authority (as at Rhodes, presumably). Thereafter it is harder to tell, for we do not know when, for example, Buondelmonti's mills were built, and their sites frequently changed rulers. Gallipoli, for instance, was Venetian 1205–35, Byzantine 1235–1354, Turkish 1354–67, Byzantine 1367–76, and finally Turkish again. Khios, where the harbour windmills were probably, as at Rhodes, built officially rather than privately, was continuously under Latin or Genoese rule from 1204, except for a brief Byzantine phase in 1326–46. Thus many of these later mills—later in our record, but not necessarily later in construction—were very likely built under non-Greek rule.

The Italians, however, who as we have seen had no tradition of windmills at home, hardly qualify as disseminators of this technology to their colonies. The Ottomans displayed even less interest. Right up to modern times, windmills in the territories they conquered were built and run by Greeks, one result being that Turkish terminology for so basic a matter as bread and its preparation includes many words of Greek origin.[130] Indeed, whether they were dominated by Italian or Turkish landlords, the great bulk of the subject populations remained Greek: intermixed, to be sure, to a greater or lesser degree with Slavonic and other blood, but still Greek. It would be they who literally built the mills, and in all probability they followed the Byzantine technical tradition, not that of their lords and masters. Thus in 1262 we find on Zakinthos, long under Norman or Venetian rule, one Nicolaos the architect: perhaps a builder, perhaps an engineer, probably a bit of both, but a Greek.[131]

THE RISE AND DIFFUSION OF THE VERTICAL MILL

The task remains of trying to make sense of the thin and scattered evidence so far assembled. All that can be done is to build up a fairly detailed model which covers the known facts and the probabilities, but which without new information can be proved neither overall nor in detail.

I have suggested that in about 1080 the horizontal windmill was brought in the wake of the Seljuq invasion from Khorasan and Sistan to the Byzantine Empire. There it took firm enough root to establish a local tradition reflected before the 1120s by the Crimean mill, around 1150 by the fantastic wind-powered palace of the *Voyage*, by the Kandia mill

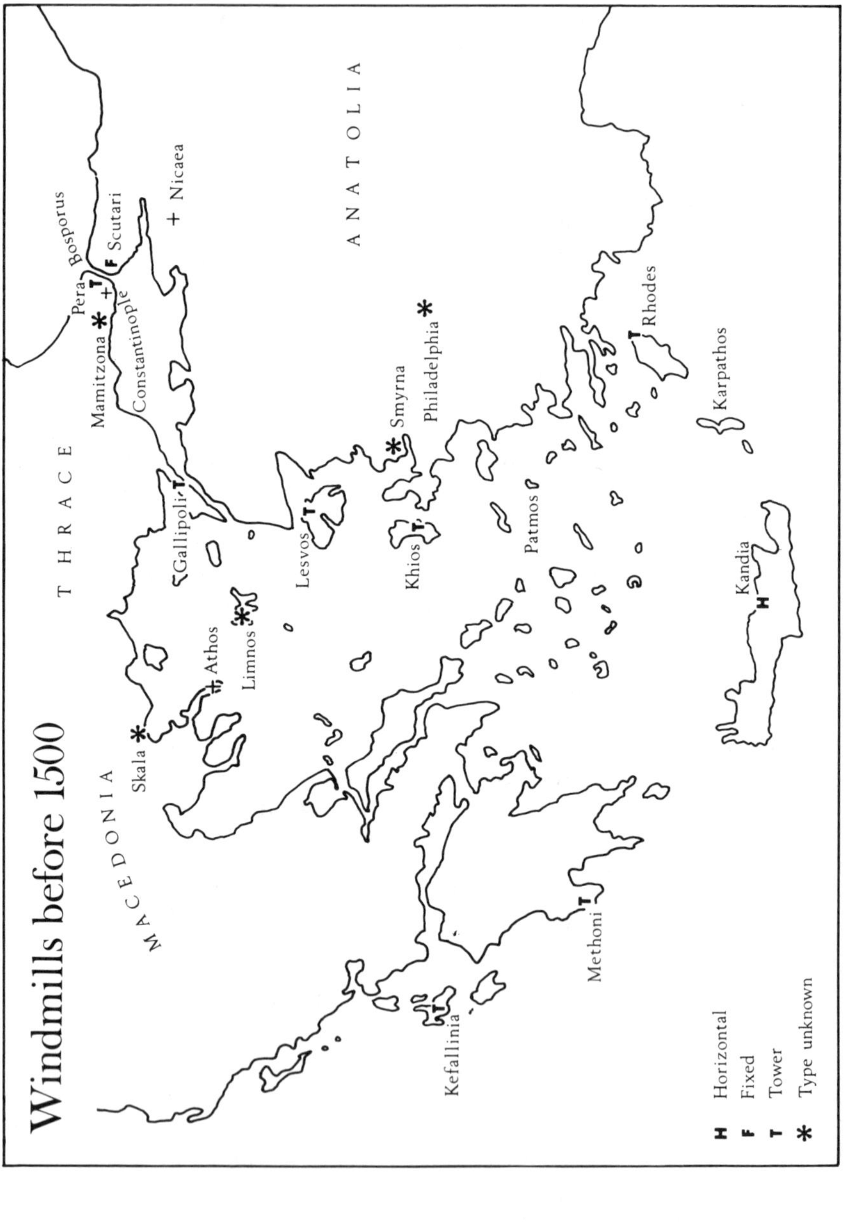

Windmills before 1500

A N A T O L I A

+ Nicaea

Bosporus

Pera
Scutari
F

Mamitzona **✳** **T**
+

Constantinople

T H R A C E

Philadelphia **✳**

Smyrna
✳

Rhodes **T**

Karpathos

Gallipoli **T**

Lesvos **T**

Khios **T**

Patmos

Kandia **H**

M A C E D O N I A

Skala **✳**

Athos **✟**
Limnos **✳**

Methoni **T**

Kefallinia

H Horizontal
F Fixed
T Tower
✳ Type unknown

H
F
T
✳

Figure 15

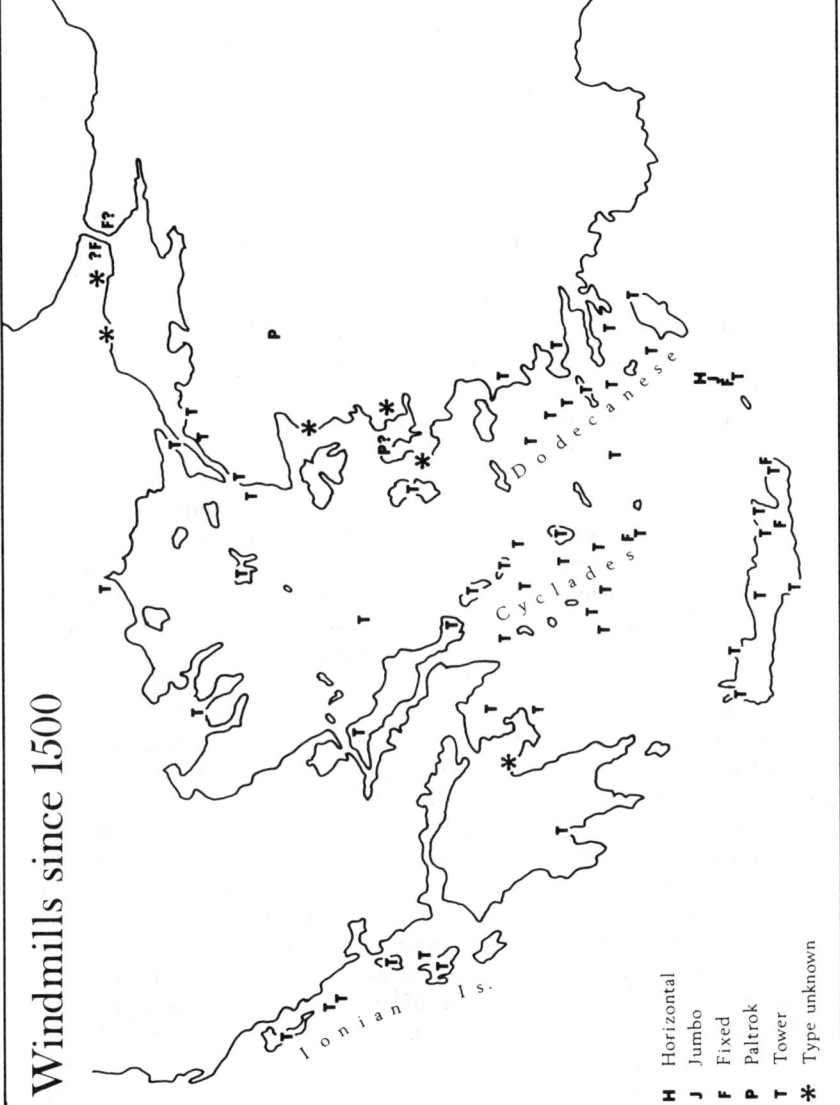

Windmills since 1500

H Horizontal
J Jumbo
F Fixed
P Paltrok
T Tower
* Type unknown

Figure 16

recorded by Grünemberg in 1486, by a rather similar mill built at Savona by the Genoese, and by the *taralis* mill of Karpathos, which closely resembled the Persian model and worked almost to the present day. To suit local patterns of winds, the design was varied from the unidirectional mills of Karpathos (and perhaps Constantinople) to the multidirectional mills of the Crimea, Kandia and Savona. Such roots as the horizontal mill put down were not as deep as in its homeland and, as happened everywhere else except Persia, it eventually succumbed to the vertical mill.

The vertical mill was an invention, I suggest, of the twelfth century which took place somewhere on the Aegean or Black Sea. What inspired it? Hardly nautical practice, for the maritime analogue of the rectangular windmill sail is a square sail mounted symmetrically on either side of the mast. This was almost or totally unknown in the Mediterranean between the ninth century at latest and the fourteenth; instead, the triangular lateen sail ruled. Nor, it would seem, did the inspiration come from watermills. True, the horizontal windmill is a close analogue of the horizontal water-mill, especially in its original form with the stones above the sails. But despite the arguments of Wailes and Notebaart,[132] the vertical windmill is not a close relative of the vertical water-wheel: the one uses an axial flow, the other a tangential. The nearest counterpart in watermill terms would be the *regolfo* or tub mill—which does employ an axial flow—rotated through a right angle.[133] This was known to the Romans[134] but only reappears after who knows what vicissitudes in sixteenth-century Spain,[135] and it would be difficult to try to relate the vertical windmill to it. We may very well, however, be dealing not with an invention but a reinvention. If the rotary fan had survived from Roman Egypt there is no Byzantine evidence for it. But the pattern for the vertical windmill already existed on paper in the form of Hero's wind-powered organ.[136]

If Hero's popularity had waned since Roman days, he was by no means ignored by the Byzantines. He was quoted extensively in a tenth-century work,[137] the oldest surviving manuscripts of his work (though not the *Pneumatica*) belong to the next century,[138] and in the twelfth he was read by the polymath John Tzetzes.[139] Under the Komnenian dynasty (1081–1180) Byzantium enjoyed a comparatively stable century and an economic and intellectual climate favourable to technical experiment. That Byzantine technology has had a generally bad press[140] is no doubt due largely to the poverty of the sources. In fact, with its inheritance of technology direct from the Roman Empire, Byzantium remained 'at least until the twelfth century'—I would myself say until the disaster of 1204—'the richest and most technically advanced state in the Mediterranean, one that provided examples for imitation'.[141] It was, for example, in this same twelfth century that the Byzantines led the way in developing that most fearsome of medieval artillery, the counterweight trebuchet.[142] Even the emperor Alexios Komnenos himself (1081–1118) had a reputation as a practical engineer.[143]

There seems no reason why some bright spark, reading his Hero in perhaps the early twelfth century, should not have recognized that here was a potential improvement to the new but inefficient horizontal windmill.

Thus Hero's device, no doubt with the number of sails modified, was married to the gearing of the vertical watermill and at last produced fruitful offspring. Alternatively, it is not inconceivable that the gearing came a little later and that, under the influence of Hero's trips, the transformation began with pounding mills: that, inspired by the Persian rice mill, the new sails were first applied to husking and crushing oats and the like, just as in some simple Russian examples.[144] Again, in view of the early use of wind power in Anatolia for lifting water, possibly some of the first vertical mills had trip-operated pumps, or buckets and chains for which gearing is not essential. Neither function requires so much power as grinding.

Vertical sails did however introduce complications because, as Hero himself had seen, they normally need to be turned to face the changing wind. If we ignore the Jumbo mill, there are three fundamental forms of vertical mill: the fixed unidirectional mill; the paltrok and its sub-species the post mill, wherein the sails and machinery are carried in a body which is turned to the wind; and the tower mill, where the machinery is contained in a fixed body and only the cap carrying the sails and first gear is rotated. Of these the fixed mill is obviously the simplest but by no means necessarily the earliest; it may have originated only later when builders realized there were places where the wind blows mainly from one direction. These being rare, it is no surprise that the fixed mill has never been widespread. Although sporadic examples are found throughout Europe, it is only in the Aegean—Thira, Crete, Karpathos and around Constantinople—that it shows any sign of following a long-standing tradition.

The paltrok

Much more fruitful was the mill with a rotatable body, normally turned by a long tailpole at the back. In its simplest form a stumpy post centred the body on the base, the bottom framing of the body simply sliding on the top framing of the base and, because of the weight it carried, generating much friction (Figure 17). In a more stable and taller version the post extended higher, with an upper centring collar and a lower compartment for storage (Figure 18).[145] Either way, this is the paltrok mill, an anachronistic name which properly belongs to a sophisticated Dutch type of much later date; but it is applied more widely in default of anything better. In using the term here, I refer only to this primitive kind.

The distribution of paltroks extends from western Asia Minor northwards via the west coast of the Black Sea to all the European republics of the late Soviet Union, finally branching out east to Omsk beyond the Urals and west to Finland and the Baltic islands. Along this broad corridor by far the most probable line of diffusion is from south to north, for the debt which Russia owes to Byzantium is immeasurable. Although there are no dates for the paltrok's journey, it is most likely to have arrived first at Kiev, then the effective capital of the Rus and the main destination of economic, political and cultural influences which travelled up the highway of the Dnieper from Byzantium. As a parallel, record survives of a water-powered saw-mill at Korsun, south of Kiev on the rapids of the River Ros. It existed by 1195, earlier than any such saw-mill known in the medieval West, and

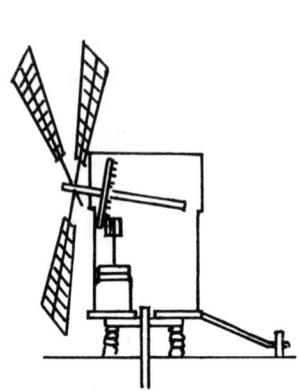
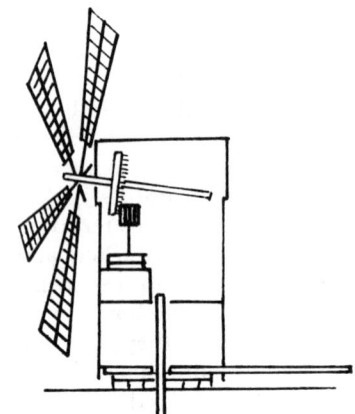

Figure 17 Romanian single-bearing paltrok (after Ruşdea). (note 145)

Figure 18 Romanian double-bearing paltrok (after Ruşdea). (note 145)

it was surely inspired by Byzantium rather than from the north.[146] During the twelfth century, however, the Cumans cut the Dnieper route, Kiev consequently lost its supremacy, and the national centre of gravity moved northwards to Novgorod and Yaroslav.[147] It was followed, I suspect, by the paltrok; since the later mills of Latvia and its neighbours were of this type it is a fair bet that a mill built at Riga in 1330 was too.[148] The paltrok, then, is essentially an eastern European mill. Elsewhere it is found only in France and Portugal (and hence in Madeira and the Azores), of which more shortly.

Given the probability of diffusion from south to north, there are three areas once within the Byzantine Empire where the paltrok is recorded (Figure 20, p. 176), which may offer a clue to its early movements. One is the Kuban,[149] the hinterland of Tmutorokan (Taman) on the east of the Sea of Azov, which was held by the Byzantines from about 1150 to the 1190s. About the early history of these paltroks we know nothing, nor can we expect to; but it is by no means impossible that they were introduced in that half century. The second area is the Dobrogea, the low-lying coastal strip of Romania and Bulgaria south of the Danube mouths, lost by Constantinople to the Bulgarians in 1186, where some of the paltroks are very small and primitive in design.[150] Here too, hard dates are non-existent: the earliest surviving reference to a windmill in Romania is dated 1585,[151] but in the poverty of local records it is scarcely surprising there is nothing earlier. Once again, it is not at all impossible that the paltrok was introduced before 1186.

The third area is Phrygia, the north-western corner of Asia Minor. Apart from the very early examples already discussed, the windmills of Asiatic Turkey were otherwise confined (as far as we know) to the Dardanelles and especially to the Aegean coast; they were operated essentially by Greeks and therefore fell into decay with the massacres and population transfers of the last two centuries; and they were tower mills (Figure 16). The exception to this rule is a cluster of paltroks, inland around Balikesir in Phrygia, of which a few are still at work, built and operated by Muhaçirs.[152] These were Muslim immigrants from the Balkans who, as the Ottoman Empire was pushed back in the last hundred years or so and replaced by Christian states, left their homelands for Turkey. Many were resettled in Phrygia.[153] The Balikesir paltroks being virtually identical to some of those in the Dobrogea, we may be certain that the design was brought in by Romanian Muslims fleeing the aftermath of the Russo-Turkish war of 1877-8. These mills are therefore something of a red herring. But were there earlier paltroks in Anatolia? A strong hint is to be found in the French chart of 1685-7 on which the windmills on the west of the Gulf of Smyrna are drawn with sufficient care to distinguish them from those on the east (Figure 7). They are not a thoughtless repetition of the normal European mill symbol of the day, a post mill with a trestle; and with a base and a door above ground level they look very much more like a paltrok than a tower mill.

Although I have found no other early evidence for the paltrok in present-day Greek or Turkish lands, I suggest that its homeland was western Anatolia and that it was the first type of windmill that could be turned to face the wind. Except on occasion for a stone base, the construction is necessarily of wood, and the coastlands of Asia Minor had plentiful timber. The date of its emergence would be in the first half of the twelfth century. The design spread, I suggest, to the Dobrogea and the Kuban while these parts were still in Byzantine hands; that is, before the twelfth century was out. This dating finds some support from the case of the western Crimea, lost to Byzantium about the 1120s. Here the standard windmill was the horizontal one and the paltrok was not apparently known. Was this because the paltrok was invented too late to reach it? If around the Aegean the paltrok was once more widespread, the tower mill, emerging during the next century, virtually swept the board and has ruled there ever since. But although the Greek tower mill exerted a mighty influence westwards it had no effect northwards through Romania to Russia, where the paltrok survived while its hypothetical ancestors further south were superseded.

The post mill
The windmill *par excellence* of medieval north-west Europe was the post mill. In principle it was the same as the paltrok, but it had a different pivot. A tall vertical post was supported on a trestle (often in early days wholly or partly buried in a mound) and extended to halfway up the body. Here the weight was transferred to a single pivot on top of the post, while a centring collar was formed by the bottom frame of the body fitting around the post. It is a type restricted for all practical purposes to north-western and central Europe. The dating for once is fairly precise. The post mill clearly existed by the middle of the thirteenth century, for the oldest illustration comes from about 1250-60 or 1270[154] and timbers from a trestle are dated by carbon-14 to 1220-80.[155] As is well known, windmills are recorded in documents from the 1180s (but not earlier[156]) in England, Normandy and Flanders, where twenty-eight are known before 1200.[157] While they may

Figure 19 Typical medieval post mill, French *c.* 1340 (after Bodleian MS 264 f. 81).

Figure 20 English post mill *c.* 1325–50 (after Smithfield Decretals) (note 159).

have been true post mills from the start, it can by no means be excluded that at first they were paltroks and that the post-pivot was a slightly later improvement. Because of the distribution, the change pretty certainly took place in western Europe, not in the Byzantine world. The reason for it was probably ease of rotation: Jespersen reckons a paltrok is more than three times harder to turn into the wind than a post mill.[158]

Two hints may confirm this ancestry. While medieval illustrations of the post mill tend to depict a fairly standard design (Figure 19), the obvious forerunner of the even more standard type of later centuries, there are a very few which differ markedly. Two scenes in the Smithfield Decretals, illuminated in England between 1325 and 1350, show a mill with a very squat body, an open framework attached to its underside, and a low-slung horizontal tailpole (Figure 20).[159] A similar arrangement is to be seen in the Holkham Bible Picture Book of much the same date.[160] Both are certainly post mills, for the top of the trestle is visible, and presumably the pivot is at the top of the framework and the centring collar at the bottom. This framework seems to correspond to the lower storage compartment of a paltrok as in Figure 18, and if it were boarded over the types could hardly be distinguished. Was this low-level arrangement of the bearings a survival from the archetype, whereas in most post mills they had already been moved up into the body proper? Secondly, there are still in Brittany a few tiny paltroks known as chandeliers, reminiscent in their single bearings of the smallest Romanian mills.[161] Is the type a survival of the earliest windmills in the West?

Who took the paltrok west? The Crusaders in Palestine are supposed to have heard of the Persian horizontal mill at second (or third or fourth) hand,

and transformed it into the European vertical mill. How much more likely that westerners saw a Byzantine paltrok with their own eyes and recognized at first hand how useful it could be. There was no lack of opportunity: pilgrims travelling to the Holy Land (though few diverted north of Rhodes), diplomats to-ing and fro-ing between the Crusader kingdoms and Constantinople, and especially Crusaders themselves. The most likely occasion, perhaps, was in 1147, when the Second Crusade under Louis VII of France and Conrad III of Germany marched from Constantinople down the coast of Asia Minor through Smyrna and Ephesus. Did some intelligent craftsman or landowner spot a windmill, grasp its potential value to his home patch in, say, France, and inspect it closely enough to construct something similar when he reached home a few years later? Some more years for experiment and a rather longer period for the idea to spread from its starting point to Yorkshire and Flanders could well bring us to 1180.[162]

The windmill is also attested[163] at much the same time—1182—near Lisbon, which had been recovered for the new kingdom of Portugal by English and Flemish crusaders in 1147. The date being too early for a tower mill and post mills being unknown here, it was presumably a paltrok, of which Portugal still has a fair number. These differ from their eastern and French cousins in that the body turns on wheels, no doubt in another attempt to reduce the friction. The idea is more likely to have come from north-western Europe than directly from Byzantine lands, for two reasons. First, Portugal's links with the north were then much stronger than with the east. Its first king, Afonso Henriques (1139–85), was a member of the House of Burgundy; many of the crusaders settled round Lisbon after 1147 and maintained contact with their homelands; and after 1153 French Cistercians were imported to improve the agriculture of the thinly populated land.[164] Second, had the paltrok travelled along the Mediterranean trade routes one might expect it to have dropped other seeds along the way, which evidently it did not. A northern source would be another sign that the paltrok preceded the post mill there.

The tower mill
The tower mill presents fewer problems. In its smallest and no doubt earliest form it was (and is) a squat cylindrical or slightly tapered masonry structure with a conical cap. Early examples in northern Europe have been largely swept away by the bigger and more sophisticated versions of later generations, but the standard windmill of the Mediterranean preserves the early form. The distribution, apart from a branch up the Vardar and Morava valleys in Yugoslavia, follows a corridor from the Aegean to Portugal via the Adriatic, Malta (imported from Rhodes in 1530 by the Knights), Sicily, the Balearics, France and Spain. In the Mediterranean, reliable evidence of date is lacking, but to the north, where the earliest tower mill on record was thought to be that built at Dover Castle in 1294–5,[165] it is now known that Jumièges Abbey in Normandy owned a mill 'called the stone mill or Hauville mill' in 1264. Distinctive among the four post mills in the parish, it has been known as 'the stone mill' ever since; it still survives, much rebuilt.[166] It is clear from documents and illustrations that

in north-western Europe the tower mill remained a considerable rarity until the end of the Middle Ages, whereas in the Aegean it was already the norm. It is therefore easier to envisage a westward diffusion than an eastward one. Moreover, the only date for a tower mill earlier than 1264 comes from the Byzantine world.

The source might well be the arid islands of the Cyclades and Dodecanese, where timber is at a premium[167] and where a more stable structure than the paltrok was needed to withstand the maritime gales. As to the date, it may be significant that there are no Aegean-type tower mills, only larger and more sophisticated affairs (including smock mills) of obviously much later date, in the Dobrogea or indeed in the whole of Russia.[168] One might tentatively deduce that the tower mill was invented after about 1190, when the Dobrogea and the Kuban ceased to be Byzantine and lost touch with technical developments in the Mediterranean, and when Kiev's close contacts with Byzantium had been cut by the Cumans. And it is not hard to believe, with the pilgrims, that the harbour windmills at Rhodes were built in 1249 and that they were tower mills even then. If so, the type appeared between about 1190 and 1249, which would allow time for it to travel west along the Mediterranean seaways and for a pilgrim or merchant or crusader to bring it directly or indirectly to France before 1264.

These—or any other—reconstructions raise large questions about the remarkably uniform spread of the paltrok northwards and of the tower mill westwards. It was not a matter of the presence or absence of seigneurial monopoly, of lords building one type and peasants the other.[169] Both on the Mediterranean and in Russia peasant mills (of either type) were in the majority,[170] whereas in north-western Europe most mills, post and tower alike, were owned by feudal lords. Some but not all of the answer lies in the availability of materials. In broad terms the Mediterranean region, though not so short of timber as is sometimes supposed, has traditionally built in stone, while Russia has built in timber. Yet the paltrok was common not only in the forested north and centre of Russia (which just includes Kiev) but also in the treeless steppes of the southern Ukraine. The likely reason is that the steppes, never part of the Kievan kingdom, were long the domain of pastoral nomads who would hardly build windmills. Only later when the land was brought under cultivation would the windmill be introduced from Kiev, the nearest source; and therefore it was the paltrok. It was also perhaps, as remarked, a matter of date: the Dobrogea and Russia, once colonized by the early paltrok, lost close contact with Byzantium before the tower mill emerged. By the time, however, that windmills came to be exported along the westward corridor from the Aegean, the paltrok there was already outmoded and the tower mill all the rage. But until much more work has been done on the early history of windmills in the receiving areas, only tentative answers can be offered to the question of these transmissions.

Sails
As already remarked, illustrations of Byzantine mills up to 1500 (and even later) all show double-sided sails with rectangular frames which possibly

carried cloths, possibly wooden boards; they also show enough multiple-sailed mills for us to be sure that these were a native feature. In 1547, moreover, the French traveller Belon, passing through Büyükçekmece, 32 km west of Constantinople, observed 'several windmills . . . and like all other windmills in Turkey they grind with eight wings or arms, and not with four as ours do'.[171] To this day, indeed, the distribution of boarded sails and of multiple sails north of the Mediterranean coincides quite closely with that of the paltrok mill, and is another indicator that the paltrok hailed from the Aegean. In north-western Europe, of course, the preference for four sails is overwhelming as far back as we can see. Yet a six-sailed mill at Framlingham in Suffolk, recorded by chance between 1270 and 1325,[172] might well mark another survival of Aegean practice, parallel to that suggested for the paltrok-like mill body.

The majority of Mediterranean tower mills too are multiple-sailed; but if we believe the illustrators of Byzantine mills (and they are unanimous) the typically Mediterranean jib sail was by 1500 not yet in use in the Aegean, or at least not in common use. Casola, an Italian churchman who had clearly never seen a windmill at close quarters before, does remark in 1494 that on the mills of Rhodes 'there is a way of regulating the amount of wind by enlarging and reducing certain sails which catch the wind'.[173] This could mean jib sails, but it could equally mean framed sails too. The rectangular-frame sail, which survived on the Aegean well into the nineteenth century[174] and is still to be seen in Malta, Sicily, the Balearics and parts of Spain, seems to represent the archetypal practice, with the jib sail only arriving later. It would be interesting to know when and where it first appeared.

Palestine

To look briefly in a different direction, it is by now accepted that the windmill was unknown in Palestine and adjacent areas until the Third Crusaders brought it in. One has, however, to doubt whether two of the mills attributed to them really existed. The story is often repeated from Ambroise, a Norman minstrel-poet, of how at the siege of Acre in 1190 the German soldiers built outside the walls 'the very first windmill that Syria had ever known', and terrified the besieged Saracens with this novelty.[175] But the *Itinerarium*, a prose chronicle which parallels Ambroise's poem and draws on the same source,[176] describes the same event rather differently:

> At that time, when the Germans built a huge milling machine for grinding foodstuffs, as the horses circled it and the rotating mill creaked, the Turks who were anxiously watching the moving mill reckoned that it was some kind of device for their destruction or for storming the city. For never before had such a donkey mill (*mola asinaria*) been seen in that land.[177]

The fact that the animal mill surely was known to the Turks need not worry us,[178] for neither Ambroise nor the chronicler could know what was in their minds; we simply have to choose between the prosaic chronicler and the romantic *jongleur*.

The great Crac des Chevaliers in Syria is supposed to have had a
Crusader windmill on one of its towers. But as Notebaart suspected[179] this
is a figment of Viollet-le-Duc's imagination.[180] The only evidence is that
one tower is called in Arabic *Borj al-tahūna*, mill tower. *Tahūna*, however,
does not specifically indicate a windmill but any kind of non-water-driven
mill.[181] Doubtless this was a horse mill. The one Crusader castle which we
know did have windmills was Safed, to the west of Galilee and begun in
1240. A contemporary wrote of it after its completion, 'There are also
twelve watermills outside the castle and many animal and wind mills inside
it, and plenty of ovens'.[182] If the Crusaders did not acquire the idea of the
windmill in Palestine, they did at least import windmills to it.

CONCLUSION

The current view tends to be that the windmill leapfrogged from Persia to
France or England, transmogrifying itself as it did so from horizontal to
vertical, and that it was brought back to the Aegean in western guise. This
scenario still cannot be disproved. But it seems to me inherently more
probable that the windmill moved in shorter stages and in more direct lines.
First, about 1080, it travelled to the Aegean where it established something
of a tradition of horizontal mills, and where it was adapted into the vertical
mill under the inspiration of Hero's toy of a thousand years before. Of the
three versions that resulted, the fixed mill stayed put. The paltrok emerged
in the first half of the twelfth century and soon began two major journeys:
one northwards via the Dobrogea and the Kuban to Russia, the other
jumping in the Crusaders' baggage to France or England where it was
quickly modified into the post mill, but not before it had spread to Portugal.
And the tower mill, evolving perhaps a century later, travelled west along
the Mediterranean and directly or indirectly north-westwards too.

In this reconstruction I find myself in broad agreement with Anders
Jespersen, who has glanced all too briefly at the same problem.[183] He
postulates a 'Quarry Trail' of stone—i.e. tower—mills from the Aegean to
Portugal and hence north as far as Scandinavia, and a 'Timber Trail' of
wooden—i.e. paltrok or post—mills from Asia Minor northwards and then
west as far as England. He ventures no dates other than assuming, without
arguing the matter, that the windmill arrived in Europe from the East in
the tenth century, which seems to me too early. I also query his placing
of the post mill at the end of the 'Timber Trail'. If the paltrok travelled
overland through Russia, Poland and Germany to reach France and
England by 1180, it would surely have had to start at an unreasonably early
date. But Jespersen's most important point—and I think he was the first
to make it[184]—is that the 'Timber Trail' and the 'Quarry Trail' part at
the Bosporus.

That is the main purpose of this paper, to underline the pivotal role
played by the Byzantine Empire in windmill history. Here, and nowhere
else in the world, all the fundamental types of windmill are found together:
the horizontal, the fixed vertical, the paltrok and the tower. This fact alone
suggests that it was the melting pot into which was fed the horizontal mill,

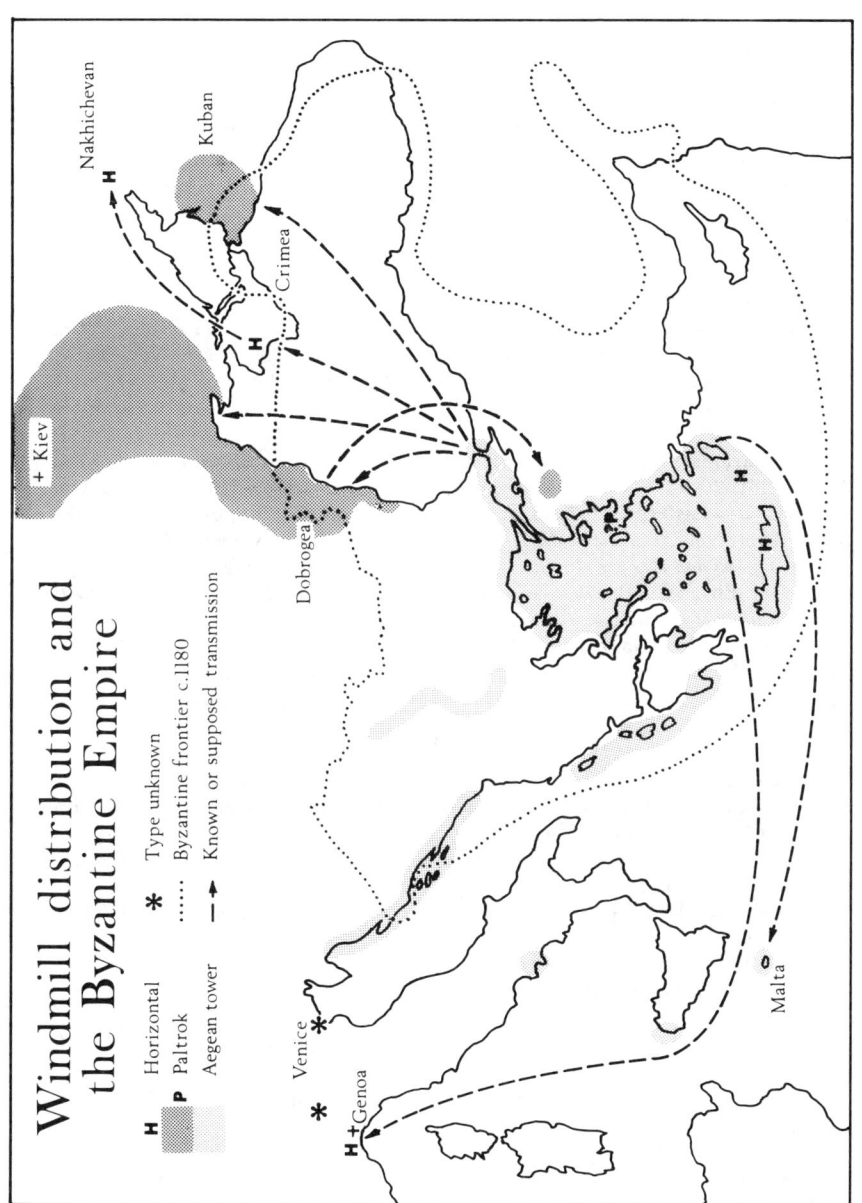

Windmill distribution and
the Byzantine Empire

H Horizontal
P Paltrok
Aegean tower

* Type unknown
.... Byzantine frontier c.1180
——▶ Known or supposed transmission

Nakhichevan
Kuban
Crimea
+ Kiev
Dobrogea
Malta
Venice *
* **H**+Genoa

Figure 21

and out of which emerged the varieties of vertical mill. Such is the core of my theory; a theory, as our knowledge stands, beyond proof and, I believe, beyond disproof. As Blaine has wisely remarked, 'Given the informational lacunae . . . such a model is useful only if used to launch further inquiry, not to close it.'[185]

Acknowledgements

I am indebted to Prof. A.A.M. Bryer, Anders Jespersen and Dr Anthony Luttrell for information, to Michael Harverson for commenting on a draft of this paper, and especially to Roy Gregory for references, much discussion, and criticizing two drafts. None of these, of course, bears any responsibility for the opinions I express.

Notes and References

1. Evliya Chelebi's curious remarks (*Narrative of Travels in Europe, Asia and Africa*, trans. Ritter Joseph von Hammer, (London, 1834, repr. New York, 1968), I.ii, 125) in the seventeenth century that 'Medina and Mecca are even now supplied only by windmills', and were so in the Prophet's time, find no support elsewhere.

2. For various views see e.g. Richard Holt, *The Mills of Medieval England* (Oxford, 1988), 20-1; Anders Jespersen, 'The Post Mill, How It Came About, a Preliminary Analysis' (paper read at 7th Symposium, 1989; publication forthcoming in *Trans. Int. Molinological Soc.*, 7); Joseph Needham, *Science and Civilization in China*, 4, *Physics and Physical Technology*, part ii, *Mechanical Engineering* (Cambridge, 1965), 564-6; Elisabeth Beazley and Michael Harverson, *Living with the Desert: Working Buildings of the Iranian Plateau* (Warminster, 1982), 93-4.

3. Jannis C. Notebaart, *Windmühlen* (The Hague and Paris, 1972), 110, 117. For all its faults, Notebaart's work is still the only international compendium of windmill typology and distribution, and I have drawn heavily on it for background material.

4. G. Demetrokalles, 'Hoi Anemomyloi ton Byzantinon', *Parnassos*, 1986, 20: 141-4. But his bibliography for recent Aegean windmills is useful.

5. A.G. Drachmann, *Ktesibios Philon and Heron*, Acta Historica Scientiarum Naturalium et Medicinalium (Copenhagen, 1948), 75-7.

6. R.J. Forbes, *Studies in Ancient Technology*, ii (Leiden, 1955), 112; R.J. Forbes, 'Power', in Charles Singer *et al.*, eds, *History of Technology*, ii (Oxford, 1956), 615.

7. Donald R. Hill, *A History of Engineering in Classical and Medieval Times* (London, 1984), 172.

8. A.G. Drachmann, 'Heron's Windmill', *Centaurus*, 1961, 7: 145-51.

9. Venice, Marcianus 516 f.184v of the thirteenth century.

10. Hero, *Pneumatica*, ed. W. Schmidt (Leipzig, 1899), 202-6.

11. Banu Musa bin Shakir, *The Book of Ingenious Devices*, trans. Donald R. Hill (Dordrecht, 1979), 222.

12. See summary of previous discussions in Drachmann, *op. cit.* (8), 150-1.

13. Cf. Hero, *Pneumatica*, 192.5, 208.5.

14. C. Dufresne Du Cange, *Glossarium ad scriptores mediae et infimae Graecitatis* (Lyon, 1688), s.v.; Alessio Da Somavera, *Tesoro della Lingua Greca-volgare ed Italiano* (Paris, 1709), s.v.

15. A.G. Paspatis, *To Chiakon Glossarion, e en Chio laloumene glossa* (Athens, 1888), 67-8.

16. Pauly-Wissowa, *Real-Encyclopädie der classischen Altertumswissenschaft*, 1

(Stuttgart, 1893–), 2182; Eustathius, *Commentarii ad Homeri Iliadem*, i (Leipzig, 1827), 222.27–30.

17. C.H. Roberts and E.G. Turner (eds), *Catalogue of the Greek and Latin Papyri in the John Rylands Library*, iv (Manchester, 1952), 627.165.

18. Galen, *Methodi Medendi* 10.14, C.G. Kühn (ed.), *Claudii Galeni Opera Omnia*, x (Leipzig, 1825), 649.

19. W.S Hüve (ed.), *Commentaria in Aristotelem Graeca*, XII.ii (Berlin, 1900), 200.16–21.

20. Needham, *op. cit.* (2), 150–5.

21. Dio Chrysostom, *Discourses*, 74.8.

22. Varro, *De Re Rustica*, III.v.17; Vitruvius, 1.6.4.

23. Various sources spell it variously—Anemodoulion or Anemoderin—but Anemodourion is the most authoritative in view of modern usage.

24. R. Janin, *Constantinople Byzantine*, Archives de l'Orient Chrétien 4 (Paris, 1950), 100–1 with references.

25. Eustathius, *Opuscula*, ed. T.L.F. Tafel (Frankfurt am Main, 1832), 109.92, 110.23–4.

26. *Etymologicon Magnum*, ed. Thomas Gaisford (Oxford, 1848), 104.4 and note; cf. also Du Cange, *op. cit.* (14), Addenda 18.

27. Zonaras, *Lexicon*, ed. J.A.H. Titmann (Leipzig, 1808), 895; Du Cange, *op. cit.* (14), s.v.; Da Somavera, *op. cit.* (14), s.v.; Ph. Koukoules, *Byzantinon Bios kai Politismos*, 2.ii (Athens, 1948), 114–5.

28. In J-P. Migne (ed.), *Patrologiae Cursus Completus, Series Graeca*, 65 (Paris, 1864), 261B.

29. Koukoules, *op. cit.* (27), 4 (1951), 276.

30. Most recently and most notably by Michael Harverson, *Persian Windmills* (Int. Molinological Soc., The Hague, 1991). For other historical notices see Notebaart, *op. cit.* (3), 210–13; *The Tārikh-e Sistān*, trans. Milton Gold, Istituto Italiano per il medio ed estremo oriente, Serie Orientale Roma xlviii (Rome, 1976), 9; and Needham, *op. cit.* (2), 559–61.

31. Harverson, *op cit.* (30), 11.

32. Needham, *op. cit.* (2), 558–61 and Fig. 689.

33. H.W. Bellew, *From the Indus to the Tigris* (London, 1874), 235.

34. G.P. Tate, *The Frontiers of Baluchistan: Travels on the Borders of Persia and Afghanistan* (London, 1909), f.p. 168. Michael Harverson (in correspondence), while not convinced by my suspicions, admits that the spokes do not look like those of a horizontal mill.

35. I owe this suggestion to Roy Gregory.

36. Illustrated in Needham, *op. cit.* (2), Fig. 697 and Rex Wailes, 'Horizontal Windmills', *Trans. Newcomen Soc.*, 1967–8, 40: 125–46, at pl. xix.

37. Needham, *op. cit.* (2), 566–7.

38. Needham, *op. cit.* (2), 390–7.

39. Al-Tabari, *Chroniques*, trans. Hermann Zotenberg, iii (Paris, 1958), 628–9.

40. Al-Masudi, *Les prairies d'or*, trans. Barbier de Meynard and Pavet de Courteille, rev. Charles Pellat, iii (Paris, 1971), 607.

41. *Motakkar ben Tāhir el-Maqdisī*, ed. and trans. Cl. Huart, v (Paris, 1916), 196–7.

42. Leone Caetani, *Annali dell'Islam*, v (Milan, 1912), 59–73.

43. H.I. Horwitz, 'Über das Aufkommen, die Entwicklung und die Verbreitung von Windrädern', *Beiträge zur Geschichte der Technik und Industrie* (1933), 22: 93ff., at 96.

44. Lynn White, *Medieval Technology and Social Change* (Oxford, 1962), 86n.

45. Caetani, *op. cit.* (42), 60.

46. Al-Tabari, *op. cit.* (39), iii, 476–80.

47. Quoted by Sir William Muir, rev. T.H. Weir, *The Caliphate, its rise, decline, and fall* (Edinburgh, 1924), 137; for the source, see p. vii.

48. Abd al-Husain Zarrinkub, 'The Arab Conquest of Iran and Its Aftermath', in R.N. Frye (ed.), *Cambridge History of Iran*, 4 (Cambridge, 1975), 1–56, at 24–7. Notebaart, *op. cit.* (3), 210 is wrong in dating the invasion to 641–2.

49. Banu Musa, *op. cit.* (11), 3.

50. Banu Musa, *op. cit.* (11), 222–3. When Hill speaks of a horizontal wheel he clearly means a horizontal axle.

51. A.D.H. Bivar, 'The History of Eastern Iran', in Ehsan Yarshater (ed.), *Cambridge History of Iran*, 3.i (Cambridge, 1983), 181–231, at 191–215.

52. For what is known of its fascinating history see W.W. Tarn, *The Greeks in Bactria and India* (Cambridge, 1938), and George Woodcock, *The Greeks in India* (London, 1976).

53. Woodcock, *op. cit.* (52), 128–31; Dio Chrysostom, *Discourses*, 32.40.

54. Woodcock, *op. cit.* (52), 129.

55. Joseph Needham, *Science and Civilization in China*, 1, *Introductory Orientations* (Cambridge, 1954), 235–6.

56. Tarn, *op. cit.* (52), 148, 309–11; Woodcock, *op. cit.* (52), 122–3; Needham, *op. cit.* (55), 234–5.

57. Philostratus, *Life of Apollonius*, iii. 27; Woodcock, *op. cit.* (52) 130–1.

58. Woodcock, *op. cit.* (52), 132–3.

59. Notebaart, *op. cit.* (3), 221.

60. Gabriel Ferrand, *Relations de voyages et textes géographiques arabes, persans et turcs relatifs à l'Extrême-Orient du VIIIe au XVIIIe siècles*, i (Paris, 1913), 194. About 1340 Ibn al-Wardi repeats al-Idrisi's description but adds that the windmills were floating ones: Ferrand, i, 418.

61. Needham, *op. cit.* (2), 564.

62. Needham, *op. cit.* (2), 561; Ahmad Y. Al-Hassan and Donald R. Hill, *Islamic Technology* (Cambridge and Paris, 1986), 55.

63. White, *op. cit.* (44), 161.

64. Bertrand Gille, 'Le moyen âge en Occident', in Maurice Daumas (ed.), *Histoire générale des techniques*, i (Paris, 1962), 431–598, at 477.

65. Edward J. Kealey, *Harvesting the air: Windmill Pioneers in Twelfth-Century England* (Woodbridge, 1987), 39–40.

66. Notebaart, *op. cit.* (3), 167–8; Sven B. Ek, 'Horizontal Windmills', *Trans. Int. Molinological Soc.*, 1965, 1: 85–100, at 90–3; Sven B. Ek, *En Skånsk Kvarn och dess Persiska Frände*, Scripta Minora 1964–5 i (Lund, 1966), 64–5.

67. Forbes, *op. cit.* (6) (1955), 116.

68. Noel Deerr, *The History of Sugar* (London, 1949), 87–92; E. Ashtor, 'Levantine Sugar industry in the Late Middle Ages—An Example of Technological Decline', *Israel Oriental Studies*, 1977, 7: 226–80, at 245–8.

69. Domenico Trevisan, in Ch. Schefer (ed.), *Le voyage d'outremer de Jean Thenaud* (Paris, 1884), 209.

70. Ek, *opp. citt.* (66), in the best discussion to date, distinguishes only two categories, elaborate and simple, not knowing of most of the Aegean examples. See also Wailes, *op. cit.* (36).

71. Eleni Limona-Trebela, *Windmills of the Aegean Sea* (Int. Molinological Soc., Reading, 1983), 1–2. Michael Harverson (in correspondence) suspects that the *taralis* mill here may have been of Italian inspiration, post-dating the Italian

occupation of the Dodecanese in 1912. This would be more convincing if there were any evidence of recent horizontal mills in Italy.

72. Limona-Trebela, *op. cit.* (71), 3–6.

73. In correspondence.

74. P.S. Pallas, *Travels through the Southern Provinces of the Russian Empire in the Years 1793 and 1794*, ii (London, 1803), 491; i (1802), 476, 478–9 and illustration p. 508 (reproduced in Wailes, *op. cit.* (36), pl. xva; Ek, *op. cit.* (66), (1966), Fig. 22; (1965), Fig. 17.07).

75. J.S., 'Power of Wind as Applied to Flour Mills', *Practical Mechanic's Journal*, 1863–4, 2nd series, 8: 231–3.

76. Space forbids documentation of the background Byzantine history given from here on. Perhaps the best starting-point is Alexander P. Kazhdan *et al.* (eds), *Oxford Dictionary of Byzantium*, 3 vols (New York and Oxford, 1991).

77. Richard Bennett and John Elton, *History of Corn Milling*, ii, *Watermills and Windmills* (London, 1899), 238. See White, *op. cit.* (44), 88 and 124, and Notebaart, *op. cit.* (3), 116–18 on the few early references to Italian windmills.

78. Marianna Koromila, *The Greeks in the Black Sea from the Bronze Age to the Early Twentieth Century* (Athens, 1991), especially 81–2, 123–4, 144–6.

79. Quoted in Ek, *op. cit.* (66) (1966), 17–18 and Fig. 7.

80. Pero Tafur, *Travels and Adventures 1435–1439*, trans. Malcolm Letts (London, 1926), 51.

81. By Bernhard von Breydenbach, *Peregrinatio in Terram Sanctam* (Mainz, 1486), and by Grünemberg (reproduced in A.L.M. Lepschy (ed.), *Viaggio in Terrasanta di Santo Brasca 1480* (Milan, 1966), pl. 15).

82. Wailes, *op. cit.* (36), 130–2; G. Marin, 'Horizontala Ventmuelilo (Ligurio)', *Geografia Revuo*, 1963, 5.5: 14, which I have not been able to see.

83. Jean-Louis G. Picherit (trans.), *The Journey of Charlemagne to Jerusalem and Constantinople* (Birmingham, Ala., 1984), lines 347–61, 365–73, 387–8; translation adapted from Picherit.

84. Margaret Schlauch, 'The Palace of Hugon de Constantinople', *Speculum*, 1932, 7:500–14.

85. J. Becker (ed.), *Die Werke Liudprands von Cremona*, Monumenta Germaniae Historica, Scriptores Rerum Germanicarum 41 (Hanover and Leipzig, 1915), *Antapodosis*, vi 5.

86. Picherit, *op. cit.* (83), vii; Jean Richard, 'Sur un passage du "Pèlerinage de Charlemagne"': le Marché de Jérusalem', *Revue belge de philologie et d'histoire*, 1965, 43: 552–5.

87. For the Seljuqs see Philip K. Hitti, *History of the Arabs*, 8th edn (London, 1968), 473–9; Osman Turan, 'Anatolia in the Period of the Seljuks and the Beyliks', in P.M. Holt *et al.*, eds, *Cambridge History of Islam*, 1 (Cambridge, 1970), 231–62; Claude Cahen, *Pre-Ottoman Turkey* (London, 1968), 155–73.

88. Speros Vryonis, 'The Question of the Byzantine Mines', *Speculum*, 1962, 37: 1–17.

89. Franz Miklosich and Jozef Müller, *Acta et Diplomata Graeca Medii Aevi Sacra et Profana*, iv (Vienna, 1871). For the fish-pond in general, 1–32, 239–47, 287–9. For the wind-pumps in particular, 17, 21, 24, 31, 242–4. For the dating of some documents see Michael Angold, *A Byzantine Government in Exile: Government and Society under the Laskarids of Nicaea (1204–1261)* (Oxford, 1975), 130, 213.

90. Anon, *Turkey*, i, Naval Intelligence Division, Geographical Handbook Series (London? 1942), 401.

91. Paul Lemerle, *L'émirat d'Aydin, Byzance et l'Occident: recherches sur 'La Geste d'Umur Pacha'* (Paris, 1957), pl. II and 43n.

92. Sophronios Eustratiadis, 'He en Philadelphia Mone tes hyperagias Theotokou tes Koteines', *Hellenika*, 1930, 3: 317–39, at 334.22–26, 34–5, 335.15–17. Beware three *anemomiliaria* that belonged to the monastery (332.9, 33). They sound like windmills, but are spelt thus, not *anemomylaria* (A. Sigalas, review of Eustratiadis, *Epeteris Hetaireias Byzantinon Spoudon*, 1931, 8: 377–81, at 381). They are probably portable boilers.

93. Anon, *op. cit.* (90), ii (1943), 157.

94. Anon, *Greece*, iii, Naval Intelligence Division, Geographical Handbook Series (London? 1945), 366–7.

95. P. Lemerle, A. Guillou, N. Svoronos and D. Papachryssanthou (eds), *Actes de Lavra* (Paris, 1970–82).

96. N. Oikonomides (ed.), *Actes de Dionysiou* (Paris, 1968), 25.12, 68, 138–9.

97. L. Petit and B. Korablev (eds), 'Actes de Chilandar', *Vizantijskij Vremennik*, 1911, 17: Supp. 1, 92.143.

98. P. Lemerle (ed.), *Actes de Kutlumus* (Paris, 1946), 24.16.

99. J. Bompaire (ed.), *Actes de Xéropotamou* (Paris, 1964), γ.4.2 (p. 235) and see pp. 230–3. Demetrokalles, *op. cit.* (4), 143 has not appreciated that it is a forgery.

100. John Haldon, 'Limnos, Monastic Holdings and the Byzantine State, ca. 1261–1453', in Anthony Bryer and Heath Lowry (eds), *Continuity and Change in Late Byzantine and Early Ottoman Society* (Birmingham and Washington, 1986), 161–215, at 175.

101. Miklosich and Müller, *op. cit.* (89), vi (1890), 56–7.

102. Miklosich and Müller, *op. cit.* (89), vi, 100, 107, 117, 388; Elisabeth Malamut, *Les îles de l'empire byzantin, VIIIe–XIIe siècle*, ii (Paris, 1988), 414–5.

103. P. Gautier (ed.), 'Le typikon du Christ Sauveur Pantokrator', *Revue des études byzantines*, 1974, 32: 1–145, at lines 999, 1006, 1049, 1255, 1258, 1266 (Constantinople 1136); P. Gautier (ed.), 'Le typikon du Sebaste Grégoire Pakourianos', *Revue des études byzantines*, 1984, 42: 5–145, at lines 392–3 (Thrace 1083); L. Petit (ed.), 'Typikon du monastère de la Kosmosoteira près d'Aenos (1152)', *Izvestija Russkogo Arheologičeskogo Instituta v Konstantinopole*, 1908, 13: 17–75, at 60.25–6 (Thrace 1152).

104. P. de Tournefort, *Relation d'un voyage du Levant* (Lyon, 1717), f.p.141 shows two or three windmills in his view of Patmos, which he visited in 1702.

105. Sp. Lampros (ed.), *Michael Akominatou Sozomena*, ii (Athens, 1880), 237.

106. J. Longnon and P. Topping, *Documents sur la régime des terres dans la Principauté de Morée au XIVe siècle* (Paris, 1969).

107. Demetrokalles, *op. cit.* (4), 142.

108. E.g. Heath W. Lowry, 'Privilege and Property in Ottoman Maçuka in the opening decades of the *Turkokratia: 1461–1553*', in Bryer and Lowry, *op. cit.* (100), 97–128; Suraiya Faroqhi, *Peasants, Dervishes and Traders in the Ottoman Empire* (Cambridge, 1986).

109. Albert Gabriel, *La cité de Rhodes MCCCX–MDXXII* (Paris, 1923), 127 for numbers.

110. Albert Gabriel, *Les remparts de Rhodes 1310–1522* (Paris, 1921), 59; Cecil Torr, *Rhodes in Modern Times* (Cambridge, 1887), 38.

111. Malcolm Letts (ed.), *The Pilgrimage of Arnold von Harff, Knight*, Hakluyt Soc. 2nd series, 94 (London, 1946), 85; Tucher quoted in Reinhold Röhricht and Heinrich Meissner (eds), *Das Reisebuch der Familie Rieter*, Bibliothek des Literarischen Vereins in Stuttgart 168 (Tübingen, 1884), 49; Felix Faber, *Evagatorium in Terrae Sanctae, Arabiae et Egypti Peregrinationem*, ed. C.D. Hassler, iii, Bibliothek des Literarischen Vereins in Stuttgart 4 (Stuttgart, 1849), 257; Thenaud, *op. cit.* (69), 126–7; Breydenbach, *op. cit.* (81); Grünemberg in Reinhold Röhricht and

Heinrich Meisner (eds), *Deutscher Pilgerreisen nach dem Heiligen Lande* (Berlin, 1880), 154.

112. Gabriel, *op. cit.* (110) 59.

113. Léon Le Grand (ed.), 'Relation du pèlerinage à Jérusalem de Nicolas de Martoni, notaire italien (1394–1395)', *Revue de l'Orient latin*, 1895, 3: 566–669, at 583–4.

114. Guillaume Caoursin, *Obsidionis Rhodie Urbis Descriptio*, BN Lat.6067 f.32, reproduced by David Chandler, *The Art of Warfare on Land* (London, 1974), 95. In the printed version (Ulm, 1496) the mills are four-sailed.

115. I have seen only the edition of 1502, with the same illustration.

116. F. Bonnardot and A. Longnon (eds), *Le saint voyage de Jherusalem du Seigneur d'Anglure*, Société des Anciens Textes Français (Paris, 1878), 95.

117. Gabriel, *op. cit.* (110), 59.

118. Émile Legrand (ed.), *Description des îles de l'archipel par Christophe Buondelmonti*, i (Paris, 1897) reproduces the maps from BN Lat.4285.

119. Georg Braun and Franz Hogenberg, *Civitates Orbis Terrarum*, iv (Cologne, *c*. 1588), 57.

120. Grünemberg in Lepschy, *op. cit.* (81), pl. 14; Ch. Schefer (ed.), *Voyage de la sainte cyté de Hierusalem* (Paris, 1882), 46; Brasca in Lepschy, 61.

121. Reproduced in e.g. Philip Sherrard, *Constantinople: Iconography of a Sacred City* (London, 1965), 13; Wolfgang Müller-Wiener, *Bildlexicon zur Topographie Istanbuls* (Tübingen, 1977), 33; and Cyril Mango, *Le développement urbain de Constantinople (IVe–VIIe siècle)*, Travaux et Mémoires Monographies 2 (Paris, 1985), Fig. 2.

122. Sebastian Münster, *Cosmographia Universalis* (Basle, 1552), 941–2.

123. Braun and Hogenberg, *op. cit.* (119), i (1572), 51. Their views were drawn from many different sources and their accuracy is correspondingly variable.

124. Hartmann Schedel, *Liber Chronicarum* (Nuremberg, 1493), cxxix.v.

125. Notebaart, *op. cit.* (3), 111; Limona-Trebela, *op. cit.* (71), 3–17.

126. Anon, *op. cit.* (90), i, 401. This is ignoring calms.

127. Evliya Chelebi, *op. cit.* (1), ii.2, 104, 124. On his reliability for Constantinople in his day, if not for more distant times and places, see Steven Runciman, *The Fall of Constantinople 1453* (Cambridge, 1990), 234.

128. Thomas Dallam in J. Theodore Bent (ed.), *Early Voyages and Travels in the Levant*, Hakluyt Society 87 (London, 1893), 57.

129. Petit and Korablev, *op. cit.* (97), 114.24.

130. Speros Vryonis, 'The Byzantine Legacy and Ottoman Forms', *Dumbarton Oaks Papers* 1969/70, 23/24: 251–308, at 291.

131. Miklosich and Müller, *op. cit.* (89), v (1887), 37.27–8.

132. Rex Wailes, *The English Windmill* (London, 1954), 150–1; Notebaart, *op. cit.* (3), 300.

133. The suggestion in Needham, *op. cit.* (2), 565 that the 'Norse' mill might have been the inspiration hardly rings true. If its blades are sometimes angled, they are only slightly so, and its water supply is hardly axial.

134. At Chemtou in Tunisia: Friedrich Rakob and Gertrud Röder, 'Die Mühle am Medjerda-Fluss', *Bild der Wissenschaft*, 1989, 26.12: 94–100.

135. Ladislao Reti, 'A Postscript to the Filarete Discussion', *Technology and Culture*, 1965, 6: 428–41, at 434 and Fig. 2; N. Garcia-Tapià, 'The Regolfo Mills of Francisco Lobarto', *Trans. Int. Molinological Soc.*, 1985, 6: 44–50.

136. Needham, *op. cit.* (2), 566–7 discusses the possible influence of Hero without totally rejecting it.

137. The Anonymus Byzantinus, sometimes called Hero of Byzantium: C. Wescher, *Poliorcétique des Grecs* (Paris, 1867), 197–279.

138. Wescher, *op. cit.* (137), xv–xxviii.

139. John Tzetzes, *Historiae*, ed. P.A.M. Leone (Naples, 1968), II.152–9, XII.965–71.

140. The fullest investigation is K. Vogel's pessimistic paper, 'Byzantine Science', in J.M. Hussey (ed.), *Cambridge Medieval History*, iv.2 (Cambridge, 1967) 264–305.

141. Kazhdan *et al.*, *op. cit.* (76), 3, 2020.

142. As I shall be arguing in a forthcoming work.

143. Anna Komnena, *Alexiad*, xi.2; Euthymius, *Panoplia Dogmatica*, Prologus, in Migne, *op. cit.* (28), 130 (1865), 20D.

144. E.E. Blomkvist, 'Krest'yanskie Postroĭki Russkikh, Ukraintsev i Belorusov', *Vostochnoslavyansky Etnografichesky Sbornik (Trudy Instituta Etnografii imeni N.N. Miklukho-Maklaya*, nov. ser. 31 (1956)), 3–458, at 324 and Fig. 74.

145. Hedriga Ruşdea, 'Windmills in Dobrogea, România. Distribution and Typology', *Trans. Int. Molinological Soc.*, 1969, 2: 447–61, at 451, 453.

146. George Vernadsky, *Kievan Russia* (New Haven and London, 1948), 300.

147. For a useful summary of Kievan Russia see W. Gordon East, *A Historical Geography of Europe*; 3rd edn (London, 1948), 217–20.

148. Notebaart, *op. cit.* (3), 177.

149. Shmuel Avitsur in *Trans. Int. Molinological Soc.*, 1969, 2: 220.

150. Ruşdea, *op. cit.* (145), 449–54; Notebaart, *op. cit.* (3), pl. 20–25.

151. Ruşdea, *op. cit.* (145), 449.

152. Richard Schultz, 'A Turkish Mill', *Int. Molinological Soc. Newsletter*, Feb. 1989, 38:5; Ewald Banse, *Die Türkei, eine moderne Geographie*, 3rd edn (Braunschweig, 1919), 70.

153. Banse, *op. cit.* (152), 15, 60; Anon, *op. cit.* (90), ii, 15–16.

154. Kealey, *op. cit.* (65), 21–3; Holt, *op. cit.* (2), 138.

155. R.J. Zeepvat, 'Post Mills and Archaeology', *Current Archaeology*, 1980, 6: 375–7. A post mill base excavated at Rieux (Oise) is claimed to be of twelfth-century date (Raymond Guilloit, *Le moulin de pierre d'Hauville* (Parc Régional de Brotonne, 1984), 21) but I do not know the evidence.

156. Kealey's attempt, *op. cit.* (65), to push the date back to 1137 has found little support: see the criticisms in Holt, *op. cit.* (2), 20n, 21n, 171–2. Nonetheless, the first cast-iron references to windmills show so wide a distribution by the mid-1180s that one has to accept that they had appeared in the West at least in the 1170s.

157. Holt, *op. cit.* (2), 171–5.

158. Jespersen, *op. cit.* (2).

159. Both illustrated in Kealey, *op. cit.* (65), Figs 23–4.

160. Kealey, *op. cit.* (65), Fig. 22.

161. M.G. Huard, Rex Wailes, H.A. Webster, 'Three Types of Windmills in Southern Brittany', *Trans. Newcomen Soc.*, 1949–51, 27: 203–10, at 209–10.

162. The name *moulin turquois* sometimes applied to post mills in Normandy may mean anything or nothing. The earliest date recorded is late—1408—and 'Turkish' was a term applied to anything oriental and exotic. See Notebaart, *op. cit.* (3), 101 with references; Anne-Marie Bautier, 'Les plus anciennes mentions de moulins hydrauliques industriels et de moulins à vent', *Bulletin philologique et historique*, 1960, 567–626, at 606; and a vague discussion in *Trans. Int. Molinological Soc.*, 1969, 2: 220.

163. A.H.R. de Oliveira Marques, *Introdução à historia da agricultura em Portugal: a questão cerealifera durante a Idade Média* (Lisbon, 1978), 195.

164. H.V. Livermore, *A New History of Portugal*, 2nd edn (Cambridge, 1976), 61–2; J.B. Trend, *Portugal* (London, 1957), 80–4.

165. John Salmon, 'A note on early tower windmills', *Journal of British Archaeological Association*, 1966, 3rd series, 29: 75.

166. Guilloit, *op. cit.* (155), 7–8, quoting a charter of Jumièges. The close similarity of the mill's south doorway to a thirteenth-century one in the sacristy of Jumièges Abbey seems to confirm its age. A mill in Hauville mentioned in 1258 may or may not be the same.

167. Although many of the larger Aegean islands had not yet been deforested, most of the Cyclades were apparently barren even then; see Anon, *op. cit.* (94), iii 216, 394, 443, 458, 516; Anon, *Dodecanese*, Naval Intelligence Division, Geography Handbook Series (London? 1941), 59.

168. Ruşdea, *op. cit.* (145), 455, 458; Notebaart, *op. cit.* (3), 178–80. Notebaart's statement (p. 53) that there were tower mills on the Bulgarian coast seems to be only a guess. In the Crimea there were both tower and smock mills (Pallas, *op. cit.* (74), ii, 491; J.S., *op. cit.* (75), 232; *Trans. Int. Molinological Soc.*, 1969, 2: 220) which like other such mills in Russia (Blomkvist, *op. cit.* (144), 322–4) were clearly of later western, not Mediterranean, type. Notebaart (p. 178) misquotes from J.S.: their height should be 22 ft., not 11 ft.

169. As Holt, *op. cit.* (2), 119–22 very plausibly argues apropos the distribution of horizontal and vertical watermills.

170. Blomkvist, *op. cit.* (144), 333.

171. Pierre Belon, *Les Observations de plusieurs Singularitez et choses memorables, trouvées en Grece, Asie, Iudée, Egypte, Arabie et autres pays etranges redigées en trois livres* (Paris, 1555), 67r.

172. Holt, *op. cit.* (2), 138–9.

173. M. Margaret Newett (ed.), *Canon Pietro Casola's Pilgrimage to Jerusalem in the Year 1494* (Manchester, 1907), 207–8.

174. Léon de Laborde, *Voyage de l'Asie Mineure* (Paris, 1838), pl. 47, at Didyma (Didim) near Miletus; the artist is careful and apparently accurate. But A.L. Castellan, *Lettres sur la Morée, l'Hellespont et Constantinople*, iii (Paris, 1820), f.p.135 illustrates mills with framed sails at Lapseki on the Dardanelles while his text talks of jib sails; these mills were turned to the wind, most unusually, by a rope pulling on the long projecting bowsprit.

175. Gaston Paris (ed.), *L'estoire de la Guerre Sainte par Ambroise* (Paris, 1871), lines 3225–32.

176. M.J. Hubert and J.L. La Monte (eds), *The Crusade of Richard Lion-Heart* (New York, 1941), 4–18.

177. William Stubbs (ed.), *Itinerarium peregrinorum et gesta regis Ricardi*, Rolls Series (London, 1864), 78–9.

178. Kealey, *op. cit.* (65), 42.

179. Notebaart, *op. cit.* (3), 228.

180. Drawing reproduced in Bennett and Elton, *op. cit.* (77), 231.

181. White, *op. cit.* (44), 161n.

182. R.B.C. Huygens, *De constructione castri Saphet. Construction et fonctions d'un château fort franc en Terre Sainte* (Amsterdam, 1981), 42.

183. Jespersen, *op. cit.* (2).

184. Hugh P. Vowles, 'An Inquiry into the Origins of the Windmill', *Trans. Newcomen Soc.*, 1930-1, 11: 1–14 postulated a line of diffusion up Viking trade routes direct from Persia through Russia to north-western Europe. This not only ignores the evidence from the Aegean and Black Sea but lacks corroboration and (to my mind) likelihood.

185. Bradford B. Blaine, 'The Enigmatic Water-Mill', in Bert S. Hall and Delno C. West, *On Pre-modern Technology and Science: Studies in Honor of Lynn White, Jr.* (Malibu, 1976), 163–76, at 167.

Faraday Reinvented: Moral Imagery and Institutional Icons in Victorian Electrical Engineering

GRAEME GOODAY

ABSTRACT

By the first centenary of his birth in 1891, telegraphic and electrical engineers had begun to create a mythology of Michael Faraday as the 'founding father' of their profession. Drawing force from the popular notions of Faraday as the 'inventor' of electromagnetic technology his name had been institutionalized in units of electrical capacitance as '*farads*' and Siemens's cable-laying ship *Faraday*.

Electrical engineers also rhetorically invoked Faraday as a moral authority in their attempt to resolve disputes over such matters as nomenclature, patents and professional leadership. Moreover, owing to the diverse characteristics widely recognized in Faraday's own career in research (scientific rigour without mathematical theory, altruistic avoidance of priority claims for technologically significant discoveries, etc.), both 'professors' and 'practical men' were able to align his posthumous identity with the interests of their factionalized positions.

This paper will document how Michael Faraday was 'reinvented' from Professor of Chemistry at the Royal Institution to archetypal representative of the Institution of Electrical Engineers in its official seal of 1899. The reasons for this extraordinary iconic elevation will be discussed in terms of Faraday's posthumous moral standing among electrical engineers in the professional controversies of the 1880s and 1890s.

> . . . when a man of genius lives a clear, calm, contemplative, serene, and dignified existence in the midst of this bustle and feverish activity, in the world and not of it, he conveys a lesson . . .
> ('The Faraday Centenary'. *The Electrician*, 19 June 1891)[1]

A little before the middle of this century Faraday brought out his great discovery, and in his own modest way anticipated some of the great

190

success which followed. He virtually gave us the dynamo . . .
(Lord Kelvin, speech at opening of power-station in Wolverhampton,
1895)[2]

Faraday . . . certainly did not know that . . . he was setting in train
a series of inventions which would ultimately revolutionize our
methods of transport and lighting, and bring about a new era in power-
using industries. Faraday's dynamo was, of course, a mere toy. It was
no more a dynamo, in the present day sense, than Watt's bubbling
kettle was a steam engine . . .
(A.G. White, *The Electrical Industry*, 1904)[3]

The image of Michael Faraday as 'founding father' of electrical engineering
was one of many identities on display during the bicentennial celebrations
of 1991. Reified in the iconography of the Institution of Electrical Engineers
(IEE) since 1899, Faraday's profile or portrait has subsequently adorned
the Institution's seals, bookplate and annual medals throughout the
twentieth century (see Figures 1 and 2).[4] His status as 'patron saint' of the

THE FIRST SEAL OF THE INSTITUTION (1899–1921)

THE SEAL SINCE 1921

Figure 1 Seals of the Institution of Electrical Engineers.

Source: R. Appleyard, *The History of the Institution of Electrical Engineers 1871–1931*
(London, 1939), 164.

Figure 2 Bookplate used in IEE Library since 1908.

Source: IEE Archives.

discipline is of course attributed to him for the results of his famous 1831 researches on electromagnetic induction, now so widely employed in the daily endeavours of electrical engineers. In this paper, however, I would like to place such issues in a broader perspective by discussing the *moral* significance of Faraday's posthumous image to late Victorian electrical engineers. In so doing I will offer a historical explanation of the exact circumstances and timing of the IEE's posthumous elevation of Faraday to iconic status, a topic which has previously received little attention from historians of technology.[5]

Elsewhere I have discussed the historical irony of considering Faraday, the solitary, qualitative and unworldly philosopher, as a representative for the heavy electrical industry, which did not come to prominence until the decade after his death in 1867.[6] As a historical phenomenon, his canonization by the IEE becomes even less self-evident when we consider that his entire career was spent as a professor of *chemistry*, moreover affiliated to the entirely independent Royal Institution. Two questions thus arise: why did the IEE need to adopt a figurehead seal in the 1890s at all? and why did it choose Faraday for this purpose in preference to distinguished figures more closely connected to the profession through membership of the IEE (or the antecedent Society of Telegraph Engineers), such as Lord Kelvin, the recently deceased John Hopkinson or the late James Clerk Maxwell?

The import of the latter question is clearer when we consider that earlier in the decade strong disagreements were visible over the legitimacy of associating Faraday with posthumously developed forms of electrical technology. During the first centenary celebrations in June 1891, two commemorative orations were given by Faraday's successors at the Royal Institution. Lord Rayleigh lectured upon the enormous significance of Faraday's researches specifically for contemporary *physics*;[7] several weeks later Sir James Dewar expounded upon the importance to chemistry of Faraday's investigations, particularly on the liquefaction of gases.[8] In reporting these events *The Times* directed attention instead to what it conceived to be the technological consequences of Faraday's work, declaring that 'the electric light, the telephone, the Atlantic cable are the direct outcome of his patient experiment.'[9] Even more bold in appropriating Faraday to the corpus of modern electrical engineers was *The Electrician*, which pronounced it 'rather discomposing to have to associate the idea of centenary with the name of FARADAY. He and his achievements belong so intensely to our own day that it is a shock to begin to memorialize him as belonging to the past.'[10]

The tension between these conflicting alignments of Faraday was evident to contemporaries. In particular the depictions of *The Times* and *The Electrician* were satirized in a cartoon published by *Punch* magazine on 27 June 1891 (Figure 3).[11] The message of the cartoon was evidently that Faraday would himself have felt somewhat 'discomposed' in encountering the ubiquitous wonders of late Victorian electro-technology, namely Edison's phonograph, Bell's telephone, Hughes' microphone, the global network of telegraph cables, and the filament bulb implied in the scene's brilliant

A SCIENTIFIC CENTENARY.

Faraday (returned). "WELL, MISS SCIENCE, I HEARTILY CONGRATULATE YOU; YOU HAVE MADE MARVELLOUS PROGRESS SINCE MY TIME!"

Figure 3 Cartoon satirizing media portrayals of Faraday during the 1891 centenary of his birth.

Source: 'A Scientific Centenary', *Punch*, 27 June 1891, 309.

illumination. All are apprehended by the 'returned' philosopher as the unfamiliar results of the 'marvellous progress' that had taken place (ironically) *since* his time. By way of an antidote to the anachronisms afoot during the centenary celebrations, Mr Punch's view of Faraday can be discerned from the test-tubes and retort flask below his right elbow: the classical icons of the laboratory chemist.

Given the manifest diversity of views upon the proper alignment of Faraday with the modern world of the 1890s, it becomes pertinent to enquire how electrical engineers succeeded in 'reinventing' Faraday as the moral archetype for *their* specialized profession. To explain the ascendancy of their particular cult of 'Faradaism'[12] I will show how certain key electrical engineers developed a moral mythology of Faraday as a crucial rallying point for the turbulently divided factions within the ranks of the IEE. In what follows I shall document the background to the inauguration

Figure 4 Sir William Siemens's 1874 double-bowed cable-laying vessel *Faraday*.

Source: H.D. Wilson, 'Notes on Submarine Cable Work: The Cable ship "Faraday" ', *The Electrician*, 1892, 30: 14.

of the IEE's seal in 1899 by analysing how Faraday's name and work were invoked in electrical debates during the preceding decades.

FARADAY RECRUITED: BRITISH TELEGRAPHY 1853–1875

At the time of his death in 1867 Faraday's legacy to posterity was unclear. Even as authoritative a figure as James Clerk Maxwell maintained that no agreement existed upon which of the philosopher's multifarious researches was his most 'significant' achievement.[13] In technological terms, at least, there was some consensus that his 1853 account of *electrostatic* induction was of greater industrial importance than his 1831 analysis of *electromagnetic* induction. Bruce Hunt has shown that British telegraph engineers and electricians successfully applied Faraday's analogical model of an underground cable as an enormous Leyden jar (capacitor) to overcome signal retardation effects encountered over long distance line working.[14] In recognition of this, the British Association's 1861 decision to christen the units of electrical measurement after the 'most eminent philosophers' eventually led to the 'Farad' being designated as the unit of capacitance.[15]

The continuing strength of Faraday's alignment with telegraphy can be seen, for example, in his widow's sanction[16] of Sir William Siemens's christening of the Atlantic cable-laying vessel as the *Faraday* (Figure 4).[17] Indeed, at the time of the ship's launch in spring 1874, Siemens lectured upon its (uniquely) purpose-built construction for telegraphic work at the Royal Institution, declaring himself 'seized with delight' at the 'opportunity to pay a tribute to the honoured name of Faraday'.[18] By contrast, an attempt by the *Telegraphic Journal* to establish an exemplary status for Faraday's researches in electromagnetism elicited little response in Britain. 'Foreigners', it complained in 1875, had been 'more demonstrative than his own countrymen' in giving the name of 'Faradaism' to his 'great discovery—magneto-electricity'.[19] Next year the journal raised the issue again, arguing that his 1831 researches were as important to the practical 'art' of telegraphy as were the 1852–4 studies in electrostatics. 'Faraday', it explained, 'determined the science of magneto-electricity' when he 'discovered the fact that currents can be produced by the motion of a wire in a magnetic field', and then Wheatstone had 'improved the art of telegraphy when he applied those currents to work a simple alphabetical dial'.[20]

Significantly, however, the message of this editorial was not only to establish the technological importance of 'magneto-electricity' but also to draw attention to Wheatstone's skill in maximizing the practical benefit of the unwordly Faraday's discovery. Lugubriously it announced that 'an expert in the art by no means implies an expert in the science, and unfortunately the two are rarely combined, especially in the telegraphic world'. As I have shown elsewhere, some success in merging these skills together was achieved by William Edward Ayrton in his scheme for training telegraphists at (what became) the Finsbury Technical College between 1879 and 1884.[21] Nevertheless, a comparable symbiosis of skills for the rising power and lighting industries was one of many problematic issues facing the new generation of electrical engineers.

FARADAY RECONSTRUED: DYNAMOS AND DYNAMISM IN THE NEW ELECTRO-TECHNOLOGY

Despite the editorial polemics of the *Telegraphic Journal*, and notwithstanding the commercial endeavours of, for example, the Belgian entrepreneur Gramme to market the self-exciting dynamo invented by Wheatstone *et al.* in 1867,[22] the technological application of Faraday's 1831 research had a relatively low public profile during the 1870s. Matters soon changed, as is revealed in Ayrton's inaugural speech as London's first Professor of Technical Physics in 1879. He adroitly traced all the burgeoning new forms of electro-technology to one point of origin:

> I will say just one word about this discovery of magneto-electricity, since on it depends all the electric lighting of the present day, and all the electric transmissions of power—that vital question of the future. . . . When Faraday first made this discovery in 1831, I have no doubt many people thought it a very pretty scientific experiment, but not likely to be of any practical use. . . . [Yet] we now see that the invention of the galvanic battery, such as we are using for producing electric light in this hall, is beginning to sink in importance before magneto-electric induction; for how do we get those strong currents to produce the electric lights of the present day on the Thames Embankment, Waterloo Bridge, the British Museum, or nearer still, in Aldersgate station? By magneto-electric induction.[23]

Ayrton went on to explain further, however, that whilst 'Faraday's principle' had also been used in other recent 'wonders' such as Hughes's microphone and Edison's telephone, the other 'ingenious principles' that had been formulated and deployed by these latter inventors were just as important as Faraday's.[24] Ayrton thereby followed professional etiquette in acknowledging the 'discoveries' of his contemporaries as being as important as those of past practitioners.

This convention was adhered to with equal fidelity by Silvanus P. Thompson in his 1883 Cantor lectures to the Society of Arts upon 'dynamo-electric machinery'. Thompson began by defining his subject as being 'those machines . . . whose action is dependent upon the principle of electro-magnetic induction, discovered by Faraday in 1831'. The main body of his discussion, though, was focused upon the developments due to subsequent electricians, starting with the first use of the term 'dynamo-electric machine' by Werner Siemens in 1867. In the first two lectures Thompson scrupulously documented the technical adaptations and variations due to twenty-seven different active practitioners including Gramme, Varley, William Thomson, Ayrton, Perry, Hopkinson, Muirhead, Edison, Ferranti, Jablochkoff, Brush and R.E.B. Crompton.[25] In this all-inclusive acknowledgement he thus articulated the co-operative labour involved in developing dynamo technology and recognized the legitimate claims of all relevant patentees.

That this degree of scrupulousness was crucial can be seen from a contemporary dispute over the priority of the first effective compound form of dynamo: a generator in which shunt and main coils were employed to make output voltage relatively independent of load. As Bowers has noted, this invention was patented in Britain by Crompton in 1882 jointly with the Austrian expatriate Gisbert Kapp.[26] Yet when Kapp discussed their compound dynamo in the pages of *The Electrician* during December 1882[27] he was rebuked by one Paget Higgs, who asserted that the American inventor Brush had got a US patent for such a dynamo in 1878 upon which Higgs himself had previously published.[28] The sharp correspondence that followed was brought to an abrupt end by a moralistic rebuke from 'W.M.', who demanded that all disputants give a 'frank acknowledgement of merit' in the work of their rivals. 'In this respect', he argued, 'no safer guide could be followed than that of Dr Faraday, whose scientific researches were too sacred to be sullied by either mere mercenary motives or factitious self-glorification.'[29]

This important normative theme of Faraday as a model of ethical behaviour was reiterated by Silvanus P. Thompson in the classic 1898 study, *Michael Faraday: His Life and Work*. In this work he depicted Faraday's examination of lighthouse illumination apparatus as a case of exemplary conduct when dealing with failures of priority acknowledgement. As Thompson waxed hagiographical about the years when Faraday was examining such apparatus for potential installation at Trinity House, he could 'calmly hear' examples described as 'Mr So-and-So's electric lights', knowing all the while that it was only 'his own discovery of magneto-electric induction' that had 'made the mechanical production of electric light possible' in the first instance.[30] Thompson's readers were left in no doubt about how Faraday's conduct was to be emulated in the priority disputes that bedevilled the later development of the dynamo.[31]

Such professional parables clearly gained their cogency directly from biographical renditions of Faraday's own unimpeachable example. The discursive power wielded by invoking Faraday's name in relation to subjects *not* explictly addressed by him or his biographers was, however, more problematic. Nevertheless, by looking at two related controversies upon such subjects—the nomenclature of electrical units and the status of practical men vs. 'professors'—we can see how electrical engineers sought to enlist the philosopher's posthumous support for their diverse sectional interests.

FARADAY REDEPLOYED: FADS AND 'FARADS' IN ELECTRICAL NOMENCLATURE

It seemed to the editor of the *Telegraphic Journal* in 1875 that 'the great inconvenience' of rival schemes of electrical nomenclature that had hitherto 'prevailed' was now 'fast dying out'; this was due 'principally' to the 'influence' of the British Association and its 'excellent reports on electrical standards'. The editor's only complaint was that certain unnamed 'professors' persisted in using the term 'intensity' to describe currents when on the authority of Faraday this could only be applied meaningfully to descriptions of potential difference.[32] He made no complaint, however, that the BAAS Committee report of 1873 had not authorized units for electrical quantity or current: none were deemed necessary since those of resistance, potential and capacitance sufficed for the purposes of most telegraphists.[33]

Matters changed, however, as the heavy electrical technologies of power and lighting emerged later in the decade. A consistent nomenclature for current became an urgent desideratum as engineers discovered to their cost the commercially disastrous consequences of misconstruing ambiguous technical specifications for electrical lighting installations. As an editorial in *The Electrician* remarked, 'even taking the point of view from which electricity is regarded as a handicraft rather than as an exact science, the inexpediency of using equivocal and ill-defined terms is frequently made manifest'. The emerging disagreements over engineering taxonomy had 'brought into prominence' the existence of 'at least two well-defined schools . . . in electrical science': the 'Scholastic or Mathematical', to which belonged the professors James Clerk Maxwell, Sir William Thomson, William Ayrton and Oliver Lodge; and the 'Physical or Engineering,' to which 'belonged at one time such men as Sturgeon and Faraday', the present representatives being Latimer Clark, Cromwell Varley, and the majority of the country's industrial practitioners.[34]

The significance of this division was elaborated further next spring when an *Electrician* editorial on 'The Nomenclature of Electrical Units' declared there had lately been no little puzzlement on the subject of Farads, Webers and Oersteds. Practitioners in the 'engineering school' of electrical practice (as well as those straddling both such as Fleeming Jenkin) had persisted in using idiosyncratic units for quantity, and current undifferentiated from that of capacitance: 'we find we have the first portion of the immortal Faraday's name doing duty both for capacity and quantity, but blending into a "Weber" in a very curious way'. For electrical quantity, Clark and others used the unit of the 'Weber', whereas Jenkin used the 'Farad' not only for capacitance but also for quantity and even current (construed as number of farads of quantity per second!). Following W.H. Preece's protocols for the Post Office telegraph service, Clark similarly used 'Weber' for both quantity and current. Others still were determined to specify currents in 'Webers per second' or even 'Oersteds'. With no little *ennui The Electrician* summarized: 'according to the authorities we have quoted, we have standing for quantity the terms "Farad" and "Weber," and for current "Farad per second", "Weber per second," "Farad", "Weber" and "Oersted" whilst "Farad" stands also for capacity.'[35]

Fortunately, the integrity of Faraday's eponymous unit and the taxonomic consensus between engineers and scientists were restored at the first international Electrical Congress in Paris in September 1881. It was here that the 'ampere' and 'coulomb' were granted a place in the electrician's lexicon for current and charge respectively. At international congresses during the subsequent decade an international consensus on units was indeed often attained by diplomatic deference to the commemorative concerns of host nations. This was particularly so when subtler new parameters such as inductance were brought into the debates upon nomenclature as the analysis of alternating current systems rose in importance later in the decade. Whilst the 'secohm' and 'quadrant' were common as units of inductance in Britain by the late 1880s, the American Institution of Electrical Engineers declared in 1891 that the 'henry' would 'honor the man who, beyond all doubt, was among the foremost to extend the knowledge thus far possessed of self and mutual inductors'.[36] Two years later, international ratification for this was secured at a major congress in Chicago[37]—a move timed to defuse imminent demands from descendants of Joseph Henry for recognition of his priority over Faraday in publishing the 'discovery' of electromagnetic induction.[38]

Although taxonomic immortalization was an effective device for relieving such international tensions, the dominant role assumed by the scientific professoriate in congress negotiations was not seen in the same diplomatic terms at home. The resolutions made at Chicago in 1893 served, in fact, to inflame once more the hostility of British 'practical men' to the abstract electrical terminology that the 'professorial' faction of Lodge, Ayrton, Thompson *et al.* were attempting to impose upon them. Foremost among the opposition was the chief electrician of the Post Office telegraph service, Sir William Preece.[39] Having previously disputed scientists' claims that 'inductance' had any relevance to the everyday concerns of electricians,[40] Preece used his presidential address to the Institution of Electrical Engineers in 1893 to attack the decisions of the Chicago congress. He specifically targeted the failure of delegates to overturn the absurd view of academic physicists that electricity was a form of matter divisible into units; in so doing he enlisted the posthumous authority of Faraday.

In almost the last investigation undertaken by Faraday on electrostatic induction in underground wires it had been Preece's 'privilege to see much of him [and] to prepare many experiments for him'.[41] It was thus 'at the feet of Faraday' that Preece had learnt that electric charge was an inductive effect of 'lines of force' acting on space and matter, not a form of matter itself. From this apparently incorrigible standpoint Preece asserted that 'electricity' means 'that form of energy which we as engineers utilise in the service of man. I do not apply it, as some physicists do, to a mere imaginary factor of this energy, sometimes called "quantity," and ever, honoured with a unit—the *coulomb*.' He suggested that this 'vague subjective unreality of the physicist' be abolished altogether. 'We can do without it. It is not wanted. It is a myth . . . The electricity of the engineer is a definite form of energy. It has one objective reality that we can generate, utilise, measure, and sell. It is even regulated by an Act of Parliament. It is moreover, "understanded of the people".'[42]

Preece's partisan wielding of Faraday's authority against the credibility of the physicists was far from unusual in the 'practical' tradition of electrical engineering. In the following year an *Electrician* editorial unfavourably contrasted the abstruseness of theoretical units introduced by the scientific professors in Chicago with the renowned simplicity of Faraday's non-mathematical work. The contrived neologisms in question included not only the (now decapitalized) 'henry' as the unit of inductance, but also the 'oersted' of reluctance, the 'weber' of magnetic flux, the 'gauss' of flux-density and the 'gilbert' of magneto-motive force. In drily mock-Tennysonian[43] vein it began: 'In spring an electrician's fancy turns lightly to thoughts of nomenclature', for 'as soon as an electrician recognises that a certain quantity is a unit', he 'yearns for a name wherewith to dub it'.[44] Such affectation was unique to the electrical branch of engineering, contended the editorial: mechanical engineers used to calibrating torsion in inch-pounds or foot-tons would 'not be grateful for a "unit" named after some celebrated wheelwright'.

Thus with reference to the 'gilbert' created at Chicago to satisfy the 'terminological hankerings' of theoreticians of alternate current systems, *The Electrician* bluntly demanded:

> Do we want a unit of magneto-motive force? Does anybody outside the lecture room use it? Is it not a clumsy analogical comparison to the objectionable expression "electro-motive force"? If something more practical than an ampere-turn is going to displace this well known quantity, we shall be interested to see it . . . FARADAY gave us the idea of a line of force, and we have given it a quantitative value; the number of lines, define them how you will, is quite sufficient for all practical purposes without any new unit of magnetic flux, and *a fortiori* we can dispense with gausses.[45]

This ironic invocation of Faraday against the scientific pretensions of the 'professors' reveals not only how valuable, but also how *malleable* was his posthumous authority in disputes over professional leadership in electrical engineering. To give a final illustration of the flexibility of Faradayan

iconography, I shall conclude this paper by showing the conciliatory use
to which it was put by the Institution of Electrical Engineers in the context
of what one late historian of the IEE has dubbed as 'the Revolution of the
Nineties'.[46]

FARADAY THE RECONCILER: PROFESSORS VS. PRACTICAL MEN IN THE IEE

The professional dominance of the 'professors' and their self-appointed
power to legislate upon matters of nomenclature was epitomized by their
high profile in the management of the IEE. Early in 1892, a large body
of its membership began to express grave dissatisfaction with the prepon-
derance of 'non-trading' engineers elected on to its council, viz. ageing
academics and wealthy amateurs. This characteristic stemmed from its
emulation of the Institution of Civil Engineers (ICE) in whose premises the
IEE had met since its foundation as the Society of Telegraph Engineers in
1871; such an arrangement enabled the older body tacitly to impose its own
constitution as a model for regulating the membership and conduct of the
younger.[47] Earlier dissatisfaction with this situation in 1890 had resulted
in a move for independence that involved the formation of a fund for
building new premises, the financial management of which legally
necessitated that the IEE have its own seal, although a formal resolution
to this effect was not made until 1895.[48]

Consideration of such moves was indeed slow until 1892, at which point
the predominance of scientific papers presented by the 'professors' on the
council at weekly meetings became intolerable for the 'very large proportion' of the membership which *The Electrician* sympathetically described
as being 'engaged in manufacture and practically in trade'. It was therefore
'quite a mistake', declared the journal, that 'trade matters' were 'very
rarely discussed' at meetings since 'many of the most influential commercial
men' were consequently 'never seen at the meetings.'[49] Remarking on the
IEE's unique status as a 'Society which should be equally representative
of Science and Engineering, Theory and Practice . . .', the editorial
encouraged the council to cultivate a 'more arithmetically accurate' consti-
tuency from the Institution's membership, in anticipation of which the
'practical men' were advised to 'forgo some of their disparagement of
theorists'.[50]

Although much of the subsequent turmoil of this period is discussed in
the standard histories of the IEE, it is necessary to supplement the
scholarship of Reader and Appleyard with an account of how the resolution
of the 'Revolution of the Nineties' related to the genesis of the Institution's
seal. A few years after commencement of the controversy over the Council's
constitution, *The Electrician* remarked that there had been a considerable
'stirring of the waters at the IEE', with plans under way not only for a
new 'independent home in a fine building'[51] but for revised statutes of
associateship. As a concession to both academic and trade interests in the
Institution (and in part to exclude amateurs), Associates would be required
to possess formal training as an electrical engineer *and* to have had

considerable experience in commercial practice.[52] This formalization of roles for both scientific and 'practical' skills in electrical engineering was effected in the form of a qualifying diploma. The regulations of this were formulated by the same committee, consisting of W.E. Ayrton and S.P. Thompson among others, then administering the design of a seal that would serve to authenticate the award of each diploma.[53]

In 1897 this committee rejected the initial choice of the 'heathen deity' Jupiter, ruler of the Olympian gods, as the proposed figurehead in the seal. The manufacturers R.E.B. Crompton and Samuel Morse successfully moved for Michael Faraday to be chosen instead.[54] Although no reasons are given for this choice in the council minutes it is clear from the foregoing that Faraday, representative not only of the scientific professors such as Ayrton and Thompson but also of the less mathematically inclined practical men such as Preece, could serve as a figure around which the various sectional interests within the Institution could unite. Not only could all electrical engineers thus agree upon him as an appropriate institutional icon, but Faraday could also publicly confer the IEE with a moral integrity that had hitherto drawn heavily upon the patronage of the ICE.

The final design of the seal adopted in 1899 was one commissioned by the ubiquitous W.E. Ayrton.[55] In the *Tribute to Michael Faraday* written for the centenary of the 1831 experiments by one of Ayrton's most eminent former students,[56] Rollo Appleyard, we see Faraday's status as a pivotal icon in the electrical engineer's history of their profession:

> For a few years there was between the 'professors' and the 'practical men' an element of raillery. Towards the end of the century, however, the support that came from the universities, and the high achievements exemplified by contributions to theory from technical men in factories and commercial laboratories, broke down prejudice, and there was established at last, on the basis of mutual appreciation, the fellowship of natural science workers for which Faraday, on neutral ground, had laboured.[57]

To conclude this paper we can note that Faraday's association with the IEE was completed by the inauguration of the library bookplate illustrated in Figure 2 in 1908,[58] and the gift of a bust of Faraday from the family of the late W.H. Preece in 1914. On receiving the sculpture on the evening of 29 October, the outgoing President William Duddell declared to the assembled members of the IEE: 'Tonight we have Faraday with us.'[59]

Acknowledgements

I should like to thank Frank James, Sophie Forgan, Geoffrey Cantor, David Gooding, Lenore Symons, Ben Marsden and Jeff Hughes for their advice and criticisms on earlier drafts of this paper. I am grateful also to the Archives of the Institution of Electrical Engineers (London) and of the Guildhall Library for permission to quote from their manuscript sources.

Research for this paper was sponsored by a British Academy Postdoctoral Fellowship and a Fellowship in Electrical History awarded by the Institution of Electrical and Electronic Engineers (New York).

Notes and References

1. 'The Faraday Centenary', *The Electrician*, 1891, 29: 186–7.

2. S.P. Thompson, *Life of Lord Kelvin* (London, 1910), 2 volumes, vol. II, 945.

3. A.G. White, *The Electrical Industry: Lighting, Traction, and Power* (London, 1904), 4.

4. Seals reproduced from R. Appleyard, *History of the Institution of Electrical Engineers 1837–1931* (London, 1939), 164; bookplate reproduced frcm specimen in IEE Archives.

5. Other images of Faraday are discussed in D. Gooding and F.A.J.L. James (eds), *Faraday Rediscovered: Essays on the Life and Work of Michael Faraday, 1791–1867* (Basingstoke, 1985); G. Cantor, *Scientist and Sandemanian* (Basingstoke, 1991). For parallel studies on the images of Edison see W. Wachhorst, *Thomas Alva Edison: An American Myth* (Cambridge, Mass. and London, 1981), and D.E. Nye, *The Invented Self: An Antibiography from documents of Thomas A. Edison* (Odense, 1983).

6. G. Gooday, 'Teaching Telegraphy and Electrotechnics in the Physics Laboratory: William Ayrton and the Creation of an Academic Space for Electrical Engineering in Britain 1873–1884', *History of Technology*, 1991, 13: 73–111. B. Bowers, *A History of Electric Light and Power* (Stevenage, 1982), 135–50; P. Dunsheath, *A History of Electrical Engineering* (London, 1962), 141–2.

7. 'The Faraday Centenary', *The Electrician*, 1891, 29: 194–6.

8. *The Times*, 30 June 1891, 13.

9. *The Times*, 18 June 1891, 9.

10. *Loc. cit.* (1).

11. 'A Scientific Centenary', *Punch*, 27 June 1891, 309.

12. Chemists' and electrical engineers' rival forms of 'Faradaism' followed an eighteenth-century pattern of divergent 'Newtonianisms', S. Schaffer, 'Natural Philosophy', in G. Rousseau and R. Porter, eds, *The Ferment of Knowledge* (Cambridge, 1980), 55–91. Cf. Wachhorst, *op. cit.* (5) and Nye, *op. cit.* (5) for the multiplicity of mythologies surrounding Edison in the USA.

13. 'James Clerk Maxwell, 'Faraday' in *Encyclopaedia Britannica*, 5th edition (London, 1875), 9: 29–31.

14. B. Hunt, 'Michael Faraday, Cable Telegraphy and the Rise of Field Theory', *History of Technology*, 1991, 13: 1–19.

15. 'Report of the Committee on Standards of Electrical Resistance'. *BAAS Report*, 1867 (Part 1): 477 and 488. From 1861–6 the 'Farad' had represented the unit of 'electrical quantity'; see 'On the Formation of Standards of Electrical Quantity and Resistance', *BAAS Report*, 1862 (Part 2): 37–38, summarized in A.C. Lynch, 'History of electrical units and early standards', *IEE Proceedings A*, 1985, 132: 564–73.

16. Sarah Faraday to Mrs C.W. Siemens, 5 February 1874, in *A Collection of letters to Sir C.W. Siemens 1823–1883* (London, 1953), 6.

17. By 1892 this ship had laid six Atlantic cables and repaired numerous others, H.D. Wilson, 'Notes on Submarine Cable Work: The Cable Ship "Faraday"', *The Electrician*, 1892, 30: 13–16.

18. Lecture reported in 'The Steamship "Faraday" and Her Appliances for Cable-Laying', *Nature*, 1874, 10: 64–5.

19. 'Faraday', *The Telegraphic Journal*, 1875, 3: 13. Its message about 'what ought to be universally called "Faradaism"' was reiterated in 'Discoveries', *The Telegraphic Journal*, 1875, 3: 97.

20. 'Art', *The Telegraphic Journal*, 1876, 4: 17.

21. Gooday, *op. cit.* (6).

22. Bowers, *op. cit.* (6), 82–90; Dunsheath, *op. cit.* (6), 108–11.

23. W.E. Ayrton, 'The Improvements Science Can Effect in Our Trades, and in the Condition of Our Workmen', *The Telegraphic Journal*, 1879, 7: 369–70.

24. *Ibid.*

25. S.P. Thompson, 'Cantor Lectures: Dynamo-Electric Machinery', *The Electrician*, 1883, 10: 187, 208, 235, 256, 270, 306, later reproduced in *Dynamo-electric Machinery* (London, 1884) and many revised editions followed.

26. Bowers, *op. cit.* (6), 93.

27. G. Kapp, 'Crompton's Compound Dynamo,' *The Electrician*, 1882, 10: 114.

28. Letter by P. Higgs, *The Electrician*, 1882, 10: 134; see Higgs, *Electric Transmission of Power: Its Present Position and Advantages* (London, 1879).

29. Letter by W.M., 'Misappropriations by Scientists', *The Electrician*, 1883, 10: 179.

30. S.P. Thompson, *Michael Faraday: His Life and Work* (London, 1898), 269.

31. E.g. the disputatious Henry Wilde's claim to have invented the 'dynamo' in 1867; see J.S. and H.G. Thompson, *Silvanus Phillips Thompson: His Life and Letters*, (London, 1920), 97–8.

32. 'Electric Terms', *The Telegraphic Journal*, 1875, 3: 73.

33. See *BAAS Report*, 1873, (Part 1): 221–5.

34. 'Electrical Terminology and Electrical Science', *The Electrician*, 1879, 3: 90–1.

35. 'The Nomenclature of Electrical Units', *Electrician*, 1880, 4: 270–1.

36. A.E. Kennelly, 'Inductance and Its Proposed Unit the Henry', *The Electrician*, 1891, 26: 267–9.

37. 'Electrical Congresses and Electrical Units', *The Electrician*, 1893, 31: 418–20.

38. 'Priority', *The Electrician*, 1894, 33: 76–7.

39. For biography see 'Sir William Henry Preece, K.C.B.', *Journal of the IEE*, 1914, 52: 793; see also E.C. Baker, *Sir William Preece F.R.S.: Engineer Extraordinary* (London, 1976), 51–4.

40. B. Hunt, 'Practice vs. Theory: The British Electrical Debate', *Isis*, 1983, 74: 341–55; D.W. Jordan, 'The Adoption of Self-Induction by Telephony, 1886–1889', *Annals of Science*, 1982, 39: 438–61.

41. Baker, *op. cit.* (39), 51–4.

42. William Preece, 'Presidential Address', *Journal of the IEE*, 1893, 22: 61–2. Inner quotation alludes to a passage in 'On speaking to the congregation' in *The Book of Common Prayer*.

43. Cf. Alfred, Lord Tennyson, 'Locksley Hall', 1837–8, line 20; see C. Ricks (ed.) *Tennyson: A Selected Edition*, (2nd edn.), (Harlow, 1987), 184.

44. 'Nomenclature and Symbols', *The Electrician*, 1894, 43: 662–3.

45. *Ibid.*

46. W.J. Reader, *A History of the Institution of Electrical Engineers 1871–1971* (London, 1987), 53–9.

47. *Ibid.*, 21; for the relationship between the 'Civils' and the IEE see R.A. Buchanan. *The Engineers: A History of the Engineering Profession in Britain, 1750–1914* (London, 1989). 90–1, 108–11.

48. *IEE Minutes of Council*, Volume 3, 1890, 268–9, IEE Archives; Appleyard, *op. cit.* (4), 164.

49. 'The Institution of Electrical Engineers', *The Electrician*, 1892, 30: 38–9.

50. *Ibid.*

51. 'The Institution Building Fund', *The Electrician*, 1896, 37: 444–5; see

earlier discussion in the *Electrical Review*, 1895, 39: 2; also the later comments of Sir Henry Mance, 'Presidential Address', *Journal of the IEE*, 1897, 26: 29.

52. 'The Institution of Electrical Engineers, *The Electrician*, 1898. 42: 198–9.

53. Appleyard, *op. cit.* (4), 164. *IEE Minutes of Council*, volume 4, 1895: 184 and 193, IEE Archives.

54. *IEE Minutes of Council*, Volume 4, 1895: 287, 289, 325; *IEE Minutes of Occasional Committees*, Volume 2, 1898: 34 and 142. See also Appleyard, *loc. cit.*

55. *Ibid.*

56. R. Appleyard, of 19 Deronda Road, Herne Hill, was an evening student in the electrical department at Finsbury Technical College in 1882–3 and 1884–5, *Register of Evening Students*, MSS 21, 973, 23 and 49, City and Guilds Institute of London Archive, Guildhall Library, London.

57. R. Appleyard, *A Tribute to Michael Faraday* (London, 1931), 85–6.

58. *IEE Minutes of the Library and Museum Committee*, I/4.5(i) (unpaginated), 1 December 1907 and 27 February, 23 March and 13 April 1908, IEE Archives.

59. *Journal of the IEE*, 1915, 53: 53–5.

PHILIPPE BRAUNSTEIN

SUMMARY

Italian legends and Silesian itineraries: prospecting for minerals in the fifteenth century

Long before the depiction of landscape became a matter of conscious artistry in Western literature there already existed utilitarian descriptions to guide the traveller or to serve as evidence in property disputes. The former, planned on a large scale, are sometimes enlivened with silhouettes or depictions of towns; the latter are small-scale, with distances marked by such features as trees or rocks or rivers or mills.

Utilitarian description and searching for minerals

When these sketches relate to mines, then a third dimension is added to such pictures in that underground workings have to be added, and, by the same token, demand an increased level of abstraction. Descriptions of this kind were developing as a genre in the German-speaking lands in the closing decades of the fifteenth century. Yet all of them have to be viewed with a certain detachment, especially when they relate to mining, for then they reflect the uncertainties inherent in prospecting for minerals. The miner had always, as one sees from the pages of Georgius Agricola, to take surface indications into account in order to assess whether a site was a good prospect or not. Typical signs were provided by the discoloration of leaves, or by patches of grass on which hoar-frost never lay. These were indicators that might lead to the discovery of the direction and dip of veins of metals, and then to the undertaking of the first surface explorations. There existed oral traditions as well as these written testimonies. Such accounts acted as a powerful incentive to search the terrain and its structures. In such situations myths of old, rich strikes inevitably blended with actuality: flooded mines and abandoned sites especially gave rise to such legends. So it is that topographical guides which undertook to direct prospectors in search of mineral riches undertook to direct them also to hidden treasures.

Mining legends: the Italians and the Germanic world

If it appeared, however, that the legends arose out of actual mining experience and had a hard factual core, that would be a reason to take them seriously, even though they talk about 'treasures', because in fact this would merely be a folkloric way of referring to mineralized zones. To the idea of work suspended with the hope of recovery in the future is added a second theme, that of the foreigner. If lost treasure was ever to be regained, it would only be with the help of outside money and technical skills not available locally. So these deserted worlds waited for the arrival of the specialists, and until they came the mountain (Berg = mine) was given over, and this is another theme, to good and evil spirits. Even the sober Agricola talks of such beings. In this legendary material the Italians (the 'Welsh') play a positive role. They are the ones who have knowledge of gold and silver, and who, one way or another, will allow the locals to grow rich in their turn. The topographical guides we are talking about here are known as the Walenbücher (Italian books), that is, collections of supposedly Italian mining knowledge. These accounts occur in a region extending from the Tyrol to Silesia, and draw upon traditions which, in the period from about 1200 to 1700, associated mining with specialists from the south called Italians or Venetians. What is astonishing in all this is that while in practice the whole world recognized that it was the Germans who were the masters of mining technology, popular tradition gave this place to Italians.

The debate over the Walenbücher

It is still far from clear how much is fact and how much is fantasy in this material. Some have sought to explain the Italian presence in the stories by their activity in the north in search of cobalt and manganese for the crystal of the glassworks of Venice. Certainly such a story is attested for Bohemia by no less an authority than Lazarus Ercker in 1574. However, none of the explanations offered so far satisfactorily explain the presence of Italians on the Silesian slopes of the Riesengebirge. We shall return later to the enigma of the presence of these Italians north of the Alps. For the moment it will be worthwhile, so to speak, to keep the Walenbücher 'whole', that is, as reflecting the migrations of specialists into regions where the local inhabitants knew nothing of the technology these strangers deployed. These 'others' may perhaps be glimpsed in the magical idea of a mirror which, looked into, revealed the interior of the mountains and God's treasures therein hidden only to those with knowledge. Such stories express a feeling of impotence among local folk once the mines had closed and the experts gone, leaving the mountains behind to shepherds and dreams. One could make progress if something like this were the case, by confronting the books with the terrain they described. Before looking at Silesia, however, it is worth taking into account a sample of this literature from Styria of *c.*1560–80. What indicates that it is a real guide to old mines and not a fiction is that 86 per cent of the place names mentioned in this account can be located on the map, although few of them are in Styria itself but lie much further east. One must ask what the object of assembling this mass of information was if it were not an archaeological compendium of someone waiting to become again an active professional. Here the Italian reference is no more than a genuflexion to an oral tradition whose meaning is no longer clearly understood. It is a very different matter when one moves eastwards to the true centre of these Italian stories: to the Silesian slopes of the Ore Mountains. Here are sites and physical remains which can be identified from the books, and here the memory of actual Italian enterprise is also clearly recalled.

The Silesian countryside and the Silesian manuscripts

Silesian mining history is, above all, a history of prospecting for gold which was to lead in the thirteenth century to something like a Californian rush. All round Reichenstein, called in 1344 'oppidum auri fodinarum', and neighbouring sites, there remained until the nineteenth century on the higher mountain slopes traces of this ancient mining. Although firm dates are wanting, the slow winding down of activity here extended from 1400 to the 1470s. Now that is precisely the time when what one may call the prototype of the Silesian Walenbücher appears, the first words of which run: 'I, Anthonius Wale, declare . . . '. All the Walenbücher can be related to this text even though the date range of these manuscripts extends from about 1500 to about 1650. This is beyond dispute even though we do not know how many links in the chain may have disappeared. An analysis of the prototype legend has found it to be composed of five texts assembled by the same hand about 1470. Together these texts represent something like an archaeological walk in the foothills of the mountains between the two Neisse rivers. Another study, this time of a manuscript copied about 1700, which combines four sources, agrees very well with the Silesian prototype preserved in the city library of Breslau (Wrocław). The links between the two series are not in doubt. The authenticity of both is also clear. The authors of these guides convince in their description because they were drawing on real knowledge. The guides in fact are rather like *aides-mémoire* to those who had understanding. But it would be naïve to think that it would be enough to follow their directions to come upon actual buried treasure. Yet the mythical is clearly an ever-present element. God distributes success as he distributes failure, and Anthonius Wale urges the prospector to prepare himself in all ways as if he were going to die. Some of these books tend to draw a veil finally over what they are purporting to reveal: nothing is to be seen all at once. Sometimes something can be before one's eyes and one will see nothing. A proverbial expression running through this whole branch of legendary German literature tells of an Italian watching peasants throwing stones to hurry up their cows, which stones, to the knowing eye, are a hundred times more valuable than the cows. The most elaborate of these guides can be read at several levels. The level closest to reality yields a view of an entirely authentic Silesian topography and history, although when the narrator feels close to his objective—gold—the Silesian prototype signals the fact by the use of diminutives, 'you will see a little hillock whence springs a rivulet in which you will find tiny flakes of gold'. This is almost like magic, with words of propitiation. One must remember also the fear that the mountains inspired in order to understand why these texts contain words of encouragement to the prospector. Climbing into the mountains one might lose more than one's nerve in the mists wreathing the Hohen Rad, or meet the Satanic Rübezahl on the icy slopes of the Abendburg. Underground the dangers were greater still. One story warns the reader of water-filled shafts and galleries in which the foul air extinguished lamps. The prospector had also better disguise himself as a herb collector if he wanted the locals to leave him undisturbed. But there is always a substrate of practical advice, such as one finds in Agricola's *De Re Metallica*. Above everything else these stories demand that attention be paid to the flow of waters, not simply surface streams whose sands might contain flakes of gold but defiles where, ear to the ground, one might hear the tinkle of invisible cascades in hidden workings. The mining areas are strewn with way markers, yet these are deceptive and difficult to interpret since each age has left its own tracks. Yet the best guides record details too authentic to be doubted.

In the corpus of Silesian Walenbücher two itineraries can be related to one another in date of composition. One leaves from Hirschberg in the direction of Schneekoppe, the other follows the route from Breslau to Silberberg. Both these itineraries were copied, or recopied, at about the same time as the Silesian prototype. Both are purged of magical references. Both descriptions correspond to the terrain they cover. One of them records the interruption of mining and loss of expertise at Reichenberg following the outbreak of the Hussite wars. Only a single old miner still lived who could point out where the lead was obtained at Schönwald, used, before 1428, to assist in the smelting of the Reichenstein silver ores rich in cobalt. This and other information establishes the value of these texts. Furthermore, it seems certain that from about 1470 these guides were actually used. It remains to be explained why the Italians are invoked as mineral prospectors and why the guides are called 'Italian books', Walenbücher.

The Silesian prototype in Breslau and Antonio 'der Wale'
Antonio 'the Italian' is the presumed author of the oldest of this, the oldest of the guides, but it has been established that an actual Richard Wales of Breslau, a Florentine, was active in Silesia between 1410 and 1443, and was employed as an agent of certain Venetian companies. Furthermore, a certain Antonio de Zane, also called 'der Wale', was authorized to exchange money in Breslau in 1410 by King Wenceslas, while one of his brothers, Leonardo, was appointed collector of St Peter's Pence in the city. Large sums of money were invested in mines, in furs, and so forth. Antonio exported copper, gold and silver to Venice. Later, Antonio 'der Wale' farmed the great Polish salt mines of Wieliczka and Bochnia before ruin overtook him in 1430. By this time the Italians in general were losing their dominant position in central and eastern Europe as wars convulsed the region. Later, when peace returned, it would be High Germans who would set about restoring mine production in Silesia and elsewhere. These efforts began after 1450, stimulated by technological invention.

As for Antonio Wales, nothing indicates that he actually mined on his own account, but the prototype Walenbuch could justifiably use him as an

authority in guiding the prospector of minerals. He had been rich and had
handled precious metals. Local memory could not believe his riches had
entirely vanished, had not been buried somewhere when danger threatened
to end the golden age of his prosperity and of gold mining alike in the
destruction of war. (Summary by Graham Hollister-Short)

SOMMAIRE

A toute époque, la maîtrise et l'exploitation du milieu naturel supposent
une capacité d'observation et d'analyse des phénomènes. Bien avant que
la littérature occidentale invente la description pittoresque du paysage,[1]
isolé dans le champ du visible par le regard d'un narrateur et ressuscité par
l'imagination ou la mémoire des lecteurs, des morceaux épars de descriptions et de représentations surgissent à des fins strictement utilitaires. Ces
types de sources découlent soit de routiers à l'usage de voyageurs, qui
éclairent des tracés sans se préoccuper des zones d'ombre qu'ils traver-
sent,[2] soit des preuves, témoignages et arguments visuels, fondant le droit
de seigneurs et de communautés comme utilisateurs du sol ou de la rente,
allégués lors de conflits sur les usages, les ressorts et les confins.[3]

Dans le premier cas, les noms de lieux ou les idéogrammes qui scandent
les tracés et les prolongent à grande échelle s'enrichissent parfois à la fin
du Moyen Age de silhouettes monumentales et d'esquisses urbaines.[4]

Dans le second cas, le paysage qui se compose sous nos yeux peut être
dominé d'un seul regard et parcouru à pied; il est délimité par un nombre
de pas, la distance d'un jet de pierre, la portée d'une arquebuse, et mesuré
par les unités théoriques qui dérivent en chaque région de gestes simples.[5]

Le paysage à ras de terre, qui se détache avec la précision nécessaire du
milieu ambiant, s'élargit aux limites marquées par des roches ou des arbres
isolés, le cours d'un ruisseau ou la berge d'un fleuve, un carrefour de
chemins, un oratoire ou un moulin.

DESCRIPTION UTILITAIRE ET PROSPECTION MINIÈRE

Étendues ou linéaires, ces descriptions fragmentées naissent d'un processus
d'abstraction, puisqu'elles supposent un point de vue d'où l'on considère
à la fois un problème à résoudre et l'espace qui le contient. A plus forte
raison, lorsqu'il s'agit de représenter des paysages souterrains, ceux de la
mine, où le parcours sinueux se conjugue avec la troisième dimension, celle
du pendage. La plus ancienne projection en surface d'un réseau souterrain
date de 1266—dans l'état actuel des connaissances—et visait à faciliter une
décision de justice.[6] Il est remarquable que ce processus d'abstraction
s'accompagnait dès le XIIIe siècle d'une grande précision des mesures.
Sans présenter toujours les mêmes garanties d'exactitude, la coupe minière
dessinée et historiée, qui fleurit dans le monde germanique à partir du
XVIe siècle, supplée par son pouvoir évocateur aux insuffisances de la
description et à l'abstraction du plan au sol.[7]

Il est une catégorie de sources descriptives médiévales et modernes que l'historiographie minière a pris en compte, sans cependant lui accorder la valeur qu'elle reconnaît aux actes de la pratique et aux illustrations topographiques. Itinéraires de surface, parfois de profondeur, dans des zones de montagne, il s'agit de documents où des données fiables voisinent avec le merveilleux, et qui sont loin d'avoir acquis le statut de preuves et témoignages parce qu'ils reflètent, sous les plumes les plus diverses et parfois les moins autorisées, les incertitudes de la prospection minière.

Sur tous les sites, avant qu'une entreprise individuelle ou collective ne s'engage, de longues recherches et des sondages se fondent sur des observations de terrain. La présentation didactique que Georg Agricola donne des raisons d'ouvrir un chantier minier aligne des considérations générales sur les facilités d'accès, les conditions climatiques, la sécurité du voisinage.[8] Sans requérir de qualités proprement techniques, l'attention à la morphologie du paysage est une condition préalable, qui suppose une acuité de vision, mais aussi des autres sens, pour détecter les indices favorables à l'entreprise. Ainsi, la plus grande importance est attachée par Georg Agricola aux eaux courantes: les sources, qu'il faut goûter, pour distinguer les six variétés possibles de solutions minérales, les ruisseaux et les fleuves dont les alluvions doivent être passées au crible.

Ensuite, c'est évidemment aux zones filonières qu'il faut consacrer le plus d'attention, en s'aidant des découvertes fortuites que la tempête ou la foundre provoquent en déracinant des arbres et en mettant à nu la roche; en suivant les miroirs de faille où s'engouffrent des cascades; en repérant les zones herbeuses où la gelée blanche ne tient pas, les alignements d'arbustes qui se distinguent par la coloration sombre de leurs tiges, et bleuâtre de leurs feuilles; enfin, en utilisant, même si sur ce point les opinions sont divisées, la baguette de coudrier. C'est à partir de ces indices que l'on doit chercher à reconnaître la direction et le pendage des filons, et pratiquer une série de fouilles au jour, avant d'entamer les premiers travaux de creusement.[9]

Il est évident que la connaissance de mines anciennes était un puissant atout pour l'analyse du paysage et de sa structure. Lorsque des privilèges de concession minière mentionnent expressément la reprise d'exploitations abandonnées, il arrive que le témoignage oral ait joué un rôle décisif dans la décision.[10] Mais des documents écrits ont également pu entretenir durablement, au-delà d'une génération de mineurs, l'espoir d'une reprise de travaux anciens. En annexe à l'édition de 1550 de sa *Cosmographie*, Sebastian Münster a tenu à publier un rapport sur les mines vosgiennes que lui avait proposé le juge territorial de Ribeaupierre, Johannes Haubensack. La renaissance contemporaine des entreprises minières dans le Val-de-Lièpvre reposait sur une longue expérience 'à peu près sur tous les versants où l'on a reconnu d'anciens puits appelés "bingen", ce qui permet aujourd'hui de voir de ses propres yeux des traces d'exploitation remontant à plus de cent ans, et attestées par quelques anciens écrits'.[11]

Que l'on renoue avec la tradition, et les témoignages se mêlent inévitablement aux légendes des succès passés. Une période d'abandon, comme en ont connu tous les sites miniers envahis par les eaux, est propice à l'efflorescence du fantastique. Or l'écrit ne relève pas toujours du témoignage, il peut être tout entier sur le versant du fantastique. C'est pourquoi, parmi les légendes que la tradition populaire a entretenues dans les zones d'exploitation minière,[12] les guides topographiques qui entendent révéler aux connaisseurs les trésors dormants des anciens sites appartiennent à un genre ambigu: ils ont été publiés par des chartistes dans des recueils de sources, par des folkloristes dans des collections de récits et légendes, par des historiens de la culture technique, comme des témoignages sur les connaissances et les représentations mentales.

LES LEGENDES MINIÈRES: LES WELSCHES DANS LE MONDE GERMANIQUE

Si le corpus légendaire né du travail minier comporte, à travers toutes les variantes du merveilleux, un noyau de références structurelles, à plus forte raison l'historiographie minière doit-elle prendre en considération quelques descriptions dont les interpolations ou l'habillage fantastique ne doivent pas dissimuler la vraie fonction: conserver, à toutes fins utiles, la trace écrite de cheminements et d'accès aux 'trésors', c'est-à-dire aux zones minéralisées. Ces itinéraires révélés partagent avec la tradition orale un mythe des origines et l'espoir d'un nouvel âge d'or.

L'interruption, parfois brutale, souvent très longue, des travaux miniers impose une première image, celle de trésors temporairement inaccessibles, que l'ingéniosité humaine, assistée de la chance, permettra un jour de retrouver. Au thème de la discontinuité et de la reprise s'ajoute un second thème, celui de l'étranger: le salut vient d'ailleurs, parce que la reprise du travail suppose que l'on franchisse des limites techniques en faisant appel à des ressources humaines et financières extérieures au site. En attendant la venue de spécialistes, la montagne—et c'est le troisième thème—est livrée à des esprits, bons ou mauvais, pâles imitations des humains, qui perpétuent dans l'entre-deux l'espoir de gain et la crainte du milieu hostile, le savoir-faire au chômage et la conscience des dangers courus. Georg Agricola lui-même consacre quelques pages imperturbables à ces ombres grimaçantes projetées sur les parois des galeries abandonnées par le souvenir des travaux souterrains.[13]

Nourries d'activités et de rêves immémoriaux, les légendes minières usent d'une syntaxe immatérielle, mais utilisent, sur des thèmes communs, le même vocabulaire que les guides topographiques à la recherche des 'trésors'. Si, dans le répertoire systématique du légendaire minier élaboré par G. Heilfurth, on considère la place réservée aux richesses cachées et à l'activité de prospection, on constate que les Italiens jouent dans les récits populaires germaniques un rôle positif: ce sont eux qui cherchent intensément les minerais d'or et d'argent et les pierres précieuses; ils transmettent directement ou indirectement leurs connaissances aux gens du pays, leur permettant de s'enrichir grâce à eux ou après leur départ définitif, fortune faite.[14]

Or, l'ensemble des guides topographiques permettant, au terme d'un parcours dans des solitudes jalonnées de repères, d'accéder à des richesses cachées est désigné par le terme de 'Walenbücher', recueils du savoir welsche.[15] Conservés dans un espace qui s'étend du Tyrol à la Silésie, ces recueils empruntent à un fonds commun de traditions qui associent, entre XIIIe et XVIIIe siècles, l'exploitation minière et la présence de spécialistes venus du Sud, désignés soit comme Welsches, soit comme Vénitiens. Malgré les lieux-communs sur la richesse de Venise, il faut sans doute assigner à cette dénomination la même valeur sémantique que l'on attribue au terme de Cahorsin. Mais il est frappant que sous des formes différentes, légendaires ou topographiques, les traditions populaires de la mine accordent aux Italiens en terre d'Empire la place que les sources de la pratique et la lexicographie minière reconnaissent à travers toute l'Europe aux spécialistes allemands.

LE DÉBAT SUR LES 'WALENBÜCHER'

La littérature scientifique sur les 'Walenbücher' n'est pas véritablement sortie d'un débat sur la fiabilité de sources contaminées par le merveilleux, et qui conduit plusieurs auteurs à jeter l'enfant avec l'eau du bain.

Un premier niveau de la discussion a porté sur l'attribution à des Italiens de guides topographiques, souvent rédigés à la première personne et s'adressant au lecteur avec familiarité. La démonstration a été faite sur le plan philologique que ces textes n'ont pu être rédigés que par des germanophones, voire par des Silésiens, connaissant parfaitement la région qu'ils décrivent, et l'on admet aisément que la référence à l'étranger vise à conférer plus d'attrait, voire plus d'autorité à des guides initiatiques.[16]

Écartant toute identification de personnages réels—et nous verrons que cette démarche est contestable—les auteurs les plus récents, comme H. Wilsdorf, tentent d'expliquer la référence italienne par une activité susceptible d'avoir intrigué les populations des zones minières au Nord des Alpes, et qui s'est poursuivie jusqu'à l'aube des temps contemporains: la recherche de minerais de cobalt et de manganèse, l'adjonction de ces substances, qui ne furent isolées et répertoriées comme métaux qu'au cours du XVIIIe siècle, étant indispensable à la production de cristal.[17] Sans doute, la verrerie de Venise a-t-elle fondé sa réputation sur des secrets chimiques, et il est satisfaisant pour l'esprit de rattacher la constance d'une référence mythique dans la tradition minière germanique à la pérennité d'un savoir-faire industriel.

Mais c'est confondre délibérément Vénitiens et Welsches, alors que les folkloristes se sont attachés à les distinguer,[18] et l'on ne peut à la fois refuser la réalité des noms italiens invoqués (vénitiens, florentins, milanais. . .) témoignant d'une conscience large d'un peuple de découvreurs,[19] et d'autre part, rapporter l'ensemble de ces témoignages à la seule recherche, attestée par une autorité de poids,[20] de substances rares que les Vénitiens auraient pu se procurer ailleurs qu'en Silésie. Encore faut-il admettre que les Italiens en question, qui demeurent pour des contemporains aussi anonymes que des agents secrets britanniques, aient camouflé leur but véritable en se faisant passer pour chercheurs d'or,[21] afin de bénéficier de l'éventuelle protection des princes lorsqu'ils s'aventuraient sur des terrains privés. L'explication paraît courte, lorsqu'on la rapproche de l'interdiction de chercher des 'trésors', concurrence déloyale vis-à-vis de concessionnaires d'exploitations minières.[22] Rétrécie à l'interprétation vénitienne et industrielle, la démythification du phénomène migratoire italien ne permet de rendre compte ni des prouesses techniques de la verrerie hors de Venise, ni de la concentration des légendes et des textes relatifs aux Italiens sur le versant silésien des Riesengebirge.

Nous reviendrons plus loin sur l'énigme philologique et culturelle de la présence italienne très au Nord des Alpes. Mais il est, par hypothèse, préférable de conserver dans toute son ampleur le schéma fécond du savoir itinérant, et de considérer les 'Walenbücher' sous l'angle d'une histoire des migrations de spécialistes, supérieurs par leurs connaissances et la qualité de leurs observations de terrain à une population locale prompte à oublier ses traditions techniques. Les légendes sur le savoir des 'autres', condensées dans le thème du miroir qui révèle l'envers de la montagne et les trésors que Dieu y a cachés à l'intention des hommes industrieux,[23] sont l'expression d'une impuissance temporaire et d'une médiation nécessaire: la fermeture d'une mine disperse les techniciens, de même que son ouverture ou sa reprise supposent connaissances et capitaux. L'entre-deux, c'est la vie ordinaire des solitudes, abandonnées aux bergers et aux rêveurs.[24]

On ne peut progresser dans l'évaluation des 'Walenbücher' qu'en considérant et en comparant leur forme et contenu, comme l'ont fait les historiens régionalistes qui, entre les deux guerres, les ont confrontés au terrain. Mais avant de nous engager sur le terrain silésien, mettons à profit l'analyse récente d'un 'Walenbüchlein' de Styrie, datant vraisemblablement des années 1560–80.[25] Passons rapidement sur l'habillage propitiatoire (calendrier favorable à la prospection, prières apotropaïques, refus de prononcer ou d'écrire le nom des minerais précieux, pour mieux les 'apprivoiser') et relevons les données topographiques qui distinguent un cheminement d'accès aux mines anciennes d'une description imaginaire.

Sur les 134 noms de lieux cités en Styrie, haute-Autriche, Carinthie et Salzbourgeois, 86% purent être reportés sur une carte, la précision des informations diminuant sensiblement d'Est en Ouest, ce qui laisse à penser qu'un noyau de notations styriennes a été entouré de compléments hétérogènes de seconde main. Quel était le but d'une semblable compilation, si ce n'est de rassembler des données pratiques, à confirmer par des promenades archéologiques? La variété des minerais énumérés, l'indication que des outils se retrouvent dans telle galerie de mine, des notations proprement historiques sur un site, tout indique que ce document constituait le compendium d'un amateur prêt à redevenir un professionnel ou à transmettre son savoir, comparable aux premiers carnets conservés, dans l'espace autrichien, de recettes métallurgiques à partir de la fin du XVe siècle. Ajoutons que la référence italienne est ici très superficielle, comme un rite obligé à une tradition orale inintelligible.

Qu'en est-il, lorsqu'on passe des marges méridionales du monde germanique vers le véritable centre de formation des légendes welsches, le versant lusacien et silésien des Monts Métallifères? On dispose d'un ensemble de données propres à faire surgir des paysages de montagne, les

'Walenbücher' eux-mêmes, les inscriptions auxquelles ces écrits font référence et qui sont en partie conservées, enfin des sites que la promenade archéologique permet d'identifier dans les 'Walenbücher'.[26] Et ici, le souvenir d'entreprises italiennes, qui débordent le cadre proprement minier, et qui ont pu nourrir la tradition topographique et légendaire.

PAYSAGES ET MANUSCRITS SILÉSIENS

Bien connue à partir de la fin du XVe siècle, grâce aux affaires des Fugger et autres firmes de haute-Allemagne investissant dans les mines et la transformation des minerais,[27] grâce aussi aux sondages effectués dans les archives de Reichenstein,[28] l'exploitation minière en Silésie est avant tout, et depuis ses origines lointaines, une histoire de chercheurs d'or. Les métaux précieux ont fortement marqué la toponymie d'une zone qui connut au XIIIe siècle un véritable rush 'à la californienne'. Dans les environs immédiats de Reichenstein, nommé en 1344 *oppidum auri fodinarum*, de Zuckmantel, de Goldberg, où le droit de Magdebourg était en 1211 conféré *hospitibus nostris de auro*, de Löwenberg, le paysage de moyenne et de haute montagne conservait au siècle dernier les traces des laveries et des mines profondes qui ont, au cours des siècles, multiplié 'pingen', haldes et bassins de décantation. Sans qu'une chronologie ait été établie, il semble que le ralentissement, puis l'interruption des travaux miniers s'étende du début du XVe siècle aux années 1470.[29] Or c'est précisément pendant cette période que serait né le prototype des 'Walenbücher' silésiens, celui qui dès ses premiers mots se réclame d'un Welsche: 'Ich Anthonius Wale vormelde gote zum lobe . . .'.

C'est en tout cas par rapport à ce texte que l'on a pu établir l'apparentement, voire la filiation de l'ensemble des 'Walenbücher' silésiens, qui s'échelonnent entre le premier tiers du XVe et le milieu du XVIIe siècle, sans méconnaître que l'usage même de ces fragiles recueils de notes manuscrites a sans doute éliminé des maillons de la chaîne. N'ont subsisté que les guides recueillis, recopiés ou publiés par des érudits des XVIIe et XVIIIe siècles; seuls le prototype et les textes qui lui sont associés ont été conservés dans une copie médiévale, déposée à la Bibliothèque de la ville de Breslau.

Analysant ce manuscrit, W.E. Peuckert[30] a décrit cinq textes d'auteurs différents, réunis par la même main vers 1470, et dont le premier touche à la région de Zittau en haute-Lusace, les deux suivants, la région de Hirschberg, le quatrième se présente comme une liste de sites d'orpaillage sur les deux versants des Monts Métallifères, et le dernier, le plus détaillé, comme un rapport sur la haute-Silésie minière, entre Reichenbach, Silberberg et Frankenstein, l'ensemble se présentant donc comme une promenade archéologique sur le piémont des Monts Métallifères entre les deux Neisse.

L'étude de K. Schneider[31] est construite à partir d'un autre manuscrit, conservé au Musée de Hohenelbe dans les Monts des Géants, datant probablement du début du XVIIIe siècle, et réunissant quatre sources: un guide attribué à un Florentin, Johannes Wale (vers 1465), déjà édité en

1619;[32] un guide attribué à Hans Man, marchand de Ratisbonne (1580), déjà édité en 1672;[33] un livret des années 1560–66 que Niclas Orler aurait obtenu de paysans vers 1650; enfin un texte composite, empruntant à la fois au 'prototype' de Breslau et au guide attribué à Johannes Wale, comportant en outre un passage original sur l'affinage de l'or.

L'apparentement entre les deux séries de textes est évident, car le guide de Johannes Wale reprend des informations qui figurent dans le 'prototype' de Breslau, tandis que les guides de Hans Man et de Niclas Orler reviennent avec précision sur la zone de Hirschberg. L'organisation spatiale et chronologique des recueils eux-mêmes, aussi hétérogènes soient-ils, obéit à un souci de représentation maîtrisée, que vient confirmer la filiation, reconstruite au XVIIIᵉ siècle, entre ces 'monuments' de l'archéologie prospective. Un manuscrit privé, conservé à Trautenau (copie datée de 1803 par un administrateur de la mine de Sebastianberg dans les Monts des Géants, d'un libelle rarissime de C.G. Lehmann paru à Leipzig en 1764),[34] rencontruit de la manière suivante la transmission écrite du

savoir sur les gîtes minéralisés de Silésie: les notes rédigées par un riche
'Vénitien' (Anthonius Wale) seraient parvenues à un riche Florentin
(Johannes Wale); un nommé Hans Gosse les aurait récupérées à la mort
de ce dernier. Or, Hans Man de Ratisbonne aurait reçu ses informations
d'un vieil Italien avec lequel il était lié en affaires, comme il l'explique lui-
même, et Niclas Orler prétend avoir reçu, un siècle plus tard, des informations écrites de paysans ignorants, qui les auraient héritées de chercheurs
d'or italiens de la région de Hirschberg, le narrateur du texte transmis par
Orler évoquant lui-même son grand-père florentin. La véracité des faits et
des enchaînements, facile à mettre en doute, impossible à vérifier, importe
ici moins que la vraisemblance du processus décrit: le quadrillage d'un
secteur minier par des promenades mentales et culturelles, superposant
de génération en génération les preuves irréfutables de la prospection
métallique et le souvenir d'un âge d'or italien.

Ce n'est pas par le biais de la relation personnelle avec leurs lecteurs—
usage du je, affirmation d'identité, rhétorique de la confidence—que ces
textes, par définition sujets à interpolations et interprétations de leurs gar-
diens successifs, persuadent de leur véracité, c'est-à-dire de la connaissance
réelle que leurs rédacteurs pouvaient avoir des zones de prospection et des
itinéraires qu'ils décrivent. Se présentant comme des aide-mémoires offerts
à qui les comprendra, ils conservent volontairement un caractère initiatique: n'y aurait-il pas quelque naïveté à croire qu'il suffit de les suivre pour
découvrir des trésors enfouis auxquels renonce l'auteur supposé de la
révélation? Que l'on peut trouver sous la mousse des pièces d'or, et dans
un puits de mine abandonné, des morceaux d'or gros comme des oeufs de
poule?[35]

L'un juxtapose l'alléchante invitation à devenir riche, et le conseil de ne
pas trop se fier aux signes marqués sur les arbres ou gravés sur les roches,
et qui servaient jusque-là de repères pour jalonner l'itinéraire;[36] l'autre
communique à son lecteur la décevante expérience personnelle du chercheur d'or, qui à sa troisième visite de terrain ne retrouve plus ses propres
traces, pour avoir entre-temps fait un mauvais usage de sa fortune.[37]

Ces entreprises risquées se jouent en effet sous le regard de Dieu, en
définitive responsable des échecs comme des succès. De même que la révélation use dans le 'prototype' de Breslau des formules solennelles du testa-
ment,[38] de même la découverte suppose un coeur pur et un esprit bien
disposé: Anthonius Wale, ou celui qui parle sous son nom, recommande
au prospecteur de 'se préparer en tout comme s'il allait mourir'.[39] Comme
un jeu de piste, dont certaines étapes, en particulier la dernière, seraient
marquées d'un carré magique substituant à la logique du cheminement une
bouffée d'irrationnel, certains 'Walenbücher' entendent occulter ce qu'ils
sont en train de révéler au public: la découverte des trésors n'est pas à la
portée du premier venu. C'est ce qu'indique tel passage sur la couleur des
pierres que l'on casse au marteau, sur la manière de reconnaître les
minerais d'argent.[40] Une expression proverbiale court à travers tout le
légendaire minier germanique, attribuée à un welsche face à des paysans:
ils jettent après leurs vaches des pierres qui valent cent fois le prix de leur
bête.[41] On ne cherche que ce que l'on connaît, on ne trouve que si l'on

vient d' 'ailleurs'. Plutôt que de dauber sur l'incohérence, voire la mal-
honnêteté des 'Walenbücher', il faut interpréter ruptures de ton et esquives
dans le sens 'substantifique', qui conserve à ces textes leur ambiguité et leur
valeur.

Les guides les plus élaborés, ceux qui ne se limitent pas à des énuméra-
tions de sites ou des formules magiques, se lisent à trois niveaux d'écriture,
sans qu'il soit nécessaire de les attribuer à plusieurs auteurs. Un premier
niveau est celui du vocabulaire initiatique qui rapproche les plus anciens
'Walenbücher' d'autres textes de la littérature de quête à travers des forêts
de symboles. Un second niveau est celui de l'observation des phénomènes,
qui révèle des gestes, des attitudes mentales, des expressions empruntées
au fonds commun de la culture technique populaire. Enfin, se dégage, dans
les textes les plus proches de la réalité vécue, un niveau de connaissances
topographiques et d'explications historiques, qui introduit à un paysage et
à un milieu minier proprement silésien.

Quand le chercheur d'or se sent près du but, le texte du 'prototype'
de Breslau l'avertit par des diminutifs, dans le style des contes de fées:
'tu verras une petite montagne (bergeleyn), d'où s'échappe un petit
ruisseau (wesserleyn), dans lequel tu trouveras de petites parcelles d'or
(goldeleyn)'.[42] On croirait entendre les formules pieuses et (ou) cabalis-
tiques, par lesquelles le prospecteur cherche à se rendre propices les forces
naturelles. Il a respecté le calendrier des jours licites,[43] où les esprits sont
impuissants; il a jeûné au pain et à l'eau, pour s'assurer la protection
divine; il fait le signe de croix avant de se mettre à l'écoute du monde
souterrain, l'oreille posée sur le sol moussu; ainsi, lorsque conseillé par
Hans Man, il monte vers le château en ruine de l'Abendburg et parvient
devant un oratoire orné d'un crucifix d'or, il n'aura aucun mal à voir les
'merveilles' de Dieu et à en prendre sa part: 'puis referme la fenêtre du
château, remets la clef à sa place, dans un trou sous la mousse, et utilise
ce bien à la gloire de Dieu . . .'.[44]

Il faut se souvenir de la peur qu'inspiraient les solitudes de haute
montagne pour apprécier les encouragements que les auteurs des guides
dispensent à leurs lecteurs: c'est dans la montée vers l'Abendburg que l'on
risque en effet de rencontrer Rübezahl, qui entre ainsi dans la littérature
de fiction. 'Ce ne sont qu'apparences, ne te retourne pas', écrit Hans
Man.[45] Près d'un siècle plus tard, une autre source revient sur les mêmes
apparitions: 'Avec tes compagnons, tu ne croiseras pas des êtres réels, ce
sont là compagnons d'épouvante ('garstige Gesellen') qui ne peuvent se
rapprocher des humains', et plus loin 'Ne te préoccupe pas, si des chimères
('Fantasey') d'esprits et de fantômes surgissent devant toi, il ne peut rien
t'arriver. . . . Recommande toi à Dieu, et passe ton chemin'.[46] Le prospecteur doit conserver le sang-froid que le choral de Luther a inspiré au
Chevalier de Dürer devant l'ennemi qui rôde ('der alt böse Feind'). Un
siècle plus tard, les esprits inoffensifs se sont démonisés, tandis que l'on
traque les sorcières. Si la quête du prospecteur est vaine malgré la qualité
des guides, c'est parce qu'elle s'inscrit dans le cadre du combat entre Dieu
et le Diable: de mauvaises gens ont, dans les solitudes, passé un pacte avec
Satan.[47]

L'observation des phénomènes naturels exige, dans ce contexte, prudence et discernement: on ne risque pas seulement de perdre son sang-froid, on risque de se perdre sans retour. Plusieurs 'Walenbücher' dispensent des conseils pratiques, qui ont la force de l'expérience personnelle. Ainsi, monter au Hohes Rad, près d'Hirschberg, est une expédition que l'on doit faire à plusieurs, en se munissant de nourriture pour huit jours, et en prenant garde à la brume: les ossements que l'on voit dans un vallon rappellent, si nécessaire, que l'on peut mourir de faim.[48] Sous terre, les dangers sont plus grands encore: tel explorateur du Weingartenloch dans le Harz signale les puits pleins d'eau, les courants d'air qui éteignent les lampes, et conseille de dérouler un filin pour retrouver la sortie des galeries.[49]

Pour pratiquer ses observations à loisir, il est enfin inutile d'attirer l'attention des populations, aussi rares que soient les rencontres dans les solitudes de haute montagne: le chercheur d'or se déguise en ramasseur de simples.[50]

Au fil des 'Walenbücher' silésiens, on trouve les conseils pratiques dont Georg Agricola fera la synthèse. D'abord, l'observation de la végétation et des roches. On est frappé par la variété des essences d'arbres citées, par l'évocation visuelle de plantes qui servent de repères, ainsi des hautes tiges, peut-être des digitales, poussant sur des éboulis d'où s'échappent des ruisselets et signalant de loin des haldes récentes,[51] monceaux de stériles, ou tas de scories métallurgiques.[52] Alerté par des couleurs ambiantes,[53] ou par la présence de roches grises ou sombres,[54] le prospecteur circule, le pic à la main, cassant les pierres pour détecter les filons minéralisés;[55] de temps à autre, il localise et signale la présence d'un 'revier', c'est-à-dire d'un gîte riche en améthystes ou en chalcédoines.[56]

Les 'Walenbücher' accordent une attention particulière à l'écoulement de l'eau dans des paysages d'altitude remodelés par d'anciens travaux miniers. Non seulement parce qu'à tout moment le prospecteur est tenté de visiter le lit des cours d'eau aurifères pour en analyser les sables, mais encore parce que des réseaux de circulation souterraine, naturels ou provoqués par le creusement des puits et des galeries, offrent des pistes sûres; on abandonne les anciens chemins de charroi[57] pour des vallons herbeux, où, sous la mousse, on entend bruisser des cascades invisibles.[58]

Les zones minières sont en outre parsemées de repères, des inscriptions gravées par des mineurs ou par d'anciens prospecteurs sur des troncs d'arbre et sur des roches.[59] Leur description ou leur transcription est parfaitement illustrée par des relevés récents. Leur utilisation n'était pas facile, car les indices et les trouvailles, échelonnées dans le temps, ont multiplié les marches d'approche qui s'entrecroisent.[60] Il ne faut donc pas les suivre comme des flèches balisant un itinéraire unique, mais comme des repères à combiner avec des éléments naturels frappants. L'auteur d'une *Chronique de Trautenau*, en Bohême orientale, démontre par son expérience personnelle que la promenade archéologique d'un amateur tourne court lorsque l'analyse du paysage se limite au constant:

L'an du Seigneur 1558, le 2° jour de novembre, mercredi après la Toussaint, moi-même Simon Hüttel, peintre, en compagnie de messire Valerius Grünberg, maître d'école à Trautenau, de Christoph Ilgnern et de Hans Teuffeln, nous avons fait un tour dans la forêt, à la recherche de la mine d'or. Nous avons trouvé de nombreuses entrées de mine, des cercles et des signes, et l'année MD2 gravée sur un bouleau avec une grande main qui indique la direction de l'Est, vers un sapin, où l'on voit un signe gravé, comme un marteau et des pointerolles.[61]

Au contraire, mis sur la bonne voie par des repères naturels ou imagés, parvenu à la zone minéralisée qu'il délimite, conjecturant la direction et le pendage des filons, le prospecteur rouvre des puits comblés,[62] ou s'engage dans l'exploration d'anciennes mines avec les précautions du professionnel. L'un des guides conservés pour la région du Harz décrit une exploration souterraine à la recherche de minerais d'argent, avec une précision qui fait revivre l'atmosphère oppressante, la promenade sportive et l'attrait du profit que le prétendu Vénitien laisse désormais à d'éventuels imitateurs.[63]

On est ici parvenu sans conteste au niveau de la réalité vécue, riche en détails qu'un inventeur en chambre ne saurait imaginer, s'il n'a lui-même été mineur ou prospecteur. Et ce n'est pas la présence d'une formule propitiatoire ou l'évocation, souvent ambiguë, de figures spectrales qui retire à de tels guides leur qualité documentaire.

Dans le *corpus* des 'Walenbücher' silésiens figurent deux itinéraires, dont l'analyse critique autorise le rapprochement chronologique.[64] L'un part de Hirschberg se dirige vers le Schneekoppe, sur les confins entre Silésie et Bohême; l'autre prend appui sur la route Breslau–Glatz et unit Silberberg et Reichenstein. Au travers de trois versions conservées, enrichies d'interpolations au cours des siècles, le premier serait construit autour d'un noyau d'informations datant des années 1460; le second, qui comporte des éléments internes de datation, aurait été recopié dès sa rédaction, vers 1470, en même temps que le 'prototype' de Breslau. Comme on peut s'en assurer aisément, les traces légendaires, les références magiques ou chrétiennes sont à peu près éliminées de ces deux textes. Que les itinéraires correspondent à la réalité des paysages de montagne qu'ils décrivent, c'est ce qu'indique d'abord leur construction qui fait passer par étapes des derniers lieux habités à la solitude des zones minières; c'est ensuite la précision oculaire,[65] qui contraste avec tant de guides embarrassés, qui recopient des indications altérées par la transmission indirecte; c'est surtout le contrôle documentaire, qui a passé au crible les noms de lieux cités, non seulement les villages traversés, mais encore les repères utilisés dans la solitude: les combes, les cours d'eau, les roches et les montagnes visées. Enfin les zones minéralisées et les lieux d'exploitation, retrouvés par des promenades archéologiques au XXᵉ siècle, comme les falaises calcaires du Spitzstein, évoquées dans le second itinéraire, avec leurs cavités souterraines.[66] Une dernière confirmation, et qui n'est pas la moindre, peut être apportée par des sources contemporaines. L'auteur donne deux informations de grand intérêt. La première évoque l'interruption des travaux d'exploitation de Reichenstein à la suite des guerres hussites, et la perte de savoir métallurgique qui s'en est suivie; seul un vieux mineur de Schönwald a été en mesure d'indiquer où se trouvait le plomb qui permettait, avant 1428, de traiter les minerais d'argent de Reichenstein, riches en cobalt; c'est une piste retrouvée pour la reprise de l'exploitation. Que la reprise soit à l'ordre du jour, c'est ce qu'indique une seconde information, relative à la mine d'Ylmenberg, qui avait été abandonnée par ses concessionnaires en discorde: le seigneur de Kaltenstein, dit le texte, favoriserait toute initiative. Or, l'évêque de Breslau qui, en 1470, se réservait expressément en ce lieu l'octroi de privilèges de prospection et de travail minier, accorde, le premier février 1473, au seigneur de Kaltenstein la liberté d'entreprise, pour lui-même et pour ses associés.[67] La véracité de l'itinéraire, et sa date probable, l'année 1473, sont donc établies. Ces preuves corroborent la valeur reconnue à des textes plus anciens, comme le 'prototype' de Breslau, puisque ce dernier fut recopié de la même main en tête du manuscrit. Mais s'il paraît désormais certain que des mineurs et des métallurgistes arpentaient vers 1470 les zones de travaux anciens, le guide à la main, il reste à expliquer pour quelle raison ces prospecteurs furent désignés comme des welsches, et leurs guides comme des 'Walenbücher'.

Même s'ils s'attachent à retrouver un noyau de vérité professionnelle sous les ornements de la fable et les rituels populaires de l'aventure, les éditeurs et critiques des 'Walenbücher' ne traitent pas leurs sources comme des documents historiques: ils cherchent à justifier par des avatars tardifs une tradition italienne, alors que, le 'prototype' de Breslau l'atteste dès ses premières lignes, la tradition repose sur l'identité d'un personnage bien connu des historiens: c'est le succès clandestin, puis public, des itinéraires de prospection et des légendes minières qui a démultiplié la silhouette et effacé le visage d'un Florentin devenu en Silésie, puis dans l'ensemble du monde germanique, l'archétype de l'étranger qui fait carrière et fortune. La notion de 'trésors'—et de 'trésors' souterrains—n'est pas étrangère à l'obscurcissement sémantique et à la représentation illusionniste d'une fortune acquise par des moyens techniques et professionnels, qui contribuèrent en leur temps à la supériorité des hommes d'affaires italiens.

Le recteur Stieff de Breslau indique en 1737 les ressorts secrets d'une interprétation collective acceptée par l'érudition, et où l'on retrouve les signes d'un contentieux culturel: combien d'Italiens, écrit-il, sont venus en pays d'Allemagne proposer expédients et 'galanteries', avant de faire commerce de vin, de produits de luxe et de bijoux, et 'après avoir amassé un grand capital s'en sont retourné en pays welsche en se moquant des naïfs Allemands qui s'étaient laissé payer de leurs beaux discours' ('ihr Geld abschwatzen lassen'). Plus généralement, on attribue la réussite économique des individus ou des firmes à la ruse italienne qui dépossède les Allemands de leurs ressources dormantes: la fable est claire, et K. Schneider indique à juste titre qu'on s'est rabattu sur une connaissance des trésors cachés faute de comprendre l'avance que possédaient les sociétés italiennes sur le terrain de l'économie de marché et des techniques administratives et bancaires.[68]

LE 'PROTOTYPE' DE BRESLAU ET ANTONIO 'DER WALE'

Si l'on s'interrogeait sur la formation et la diffusion des légendes, il suffirait de constater qu' 'Anthonius Wale', auteur présumé du plus ancien des itinéraires connus, n'a pas fini sa course historique, puisqu'il est récemment devenu, dans un ouvrage de base sur la Silésie, 'Richard Wales', résidant à Breslau.[69] Or au siècle dernier, D. Stobbe avait réuni sur la carrière silésienne de ce Florentin entre 1410 et 1443 des informations que des sources vénitiennes viennent confirmer.[70]

Les entreprises d'Antonio di Zane de' Ricci, apparenté aux Médicis, doivent être replacées dans le cadre d'une mainmise par des sociétés gênoises, puis florentines et vénitiennes sur les postes-clés de l'économie nationale du royaume de Pologne et sur l'exploitation d'un réseau commercial unissant Cracovie et Breslau à Venise. Les agents de ces sociétés disposaient en effet à Venise de sommes importantes: la collecte des fonds apostoliques dans l'Europe du Nord-Est permettait l'achat en Pologne et en Silésie de marchandises et de métaux monétaires, dont les sociétés contrôlaient l'exportation, et la revente à Venise de ces produits indispensables au trafic d'Orient.[71] Antonio de Zane, dit 'der Wale', reçut en 1410 du roi Wenceslas l'autorisation de s'installer comme changeur à Breslau; l'un de ses frères, Leonardo, est dans la même ville *collector denarii s. Petri*, un autre, Guido, apparaît quelques années plus tard comme changeur à Cracovie, le troisième, Michele, dirige le siège de Venise, et le quatrième, Bernardo, est demeuré à Florence. Les affaires traitées par le réseau Ricci et ses amis portent sur le commerce du sel et du bétail, des fourrures et de la cochenille polonaise, et associent les mouvements de fonds et le commerce des métaux monétaires: associé au marchand Johannes Bank de Breslau, Antonio 'der Wale' exporte vers Venise du cuivre, de l'argent et de l'or; on le saisit confiant à un Viennois trois morceaux d'or, et chargeant cet ami d'affaires de se renseigner sur le nombre de pièces monnayées qu'on pourrait en tirer en Hongrie ou à Venise.[72]

Propriétaire d'une maison à Breslau depuis 1419, exploitant de pêcheries, Antonio 'der Wale' s'était depuis 1425 installé en Pologne: il avait en effet pris à ferme, en compagnie de son frère Leonardo, les salines de Wieliczka et de Bochnia, et était devenu un des principaux gestionnaires du royaume.[73] Ce sont des problèmes de trésorerie qui le perdirent: sommé en 1430 de verser à la Couronne les 1500 florins qu'il lui devait et incapable de faire face à ses obligations, il fut déchu de sa charge, emprisonné en 1432, temporairement rétabli à la direction des salines sous le contrôle d'un associé polonais, mais dût finalement quitter la Pologne en 1435 en abandonnant tous ses biens. On le retrouve à Breslau où il se refit, puisqu'il était en 1439 consul de la ville; mais en 1443, il fut expulsé du conseil et banni, et l'on perd définitivement sa trace.[74]

Tandis que le roi de Pologne faisait prononcer le séquestre de ses propriétés, c'est un Allemand qui succédait à Albizzo de' Medici, son cousin, à la tête de l'Hôtel des Monnaies de Cracovie.[75]

Plus jamais les Italiens ne retrouvèrent les positions qu'ils avaient occupées au XIVe siècle dans l'ensemble de l'Europe centrale et orientale.

Evincés, ils furent remplacés dans la haute administration, les grandes entreprises et les échanges à longue distance par leurs concurrents haut-allemands qui, avec l'appui de la maison de Luxembourg, étaient devenus au début du XVᵉ siècle les intermédiaires exclusifs entre l'Europe du Nord et le monde méditerranéen.[76] Cependant, depuis une génération, des techniciens appartenant au milieu nurembergeois des affaires tentaient de restaurer la prospérité minière en Silésie, comme en Hongrie ou dans le Harz.[77] Retardée par les guerres hussites, la reprise fut stimulée à partir du milieu du XVᵉ siècle par le marché et par l'innovation technique.[78] Il est significatif que dans la principale région aurifère de Silésie, la zone de Reichenstein, la référence du 'Walenbuch' vers 1470 ne soit plus l'Italien, mais un Allemand, nommé Procopius Hoberg. Ce personnage n'a pas été identifié, si tant est qu'il ait existé; mais son prénom nous met sur la piste de Prague et, par conséquent, de Nuremberg.[79] A ce titre, il est parfaitement illustratif des changements conjoncturels survenus dans la production et le commerce des métaux précieux.

Rien n'indique assurément qu'Antonio 'der Wale' ait ajouté à ses activités marchandes et financières des entreprises minières pour son propre compte. A tout le moins, le commerce de l'or, étroitement lié aux mouvements de fonds vers l'Italie, avait été un secteur-clé de ses occupations silésiennes. Le 'prototype' des 'Walenbücher' peut à juste titre se réclamer de son autorité fondatrice, même s'il est exclu pour des raisons philologiques que le texte lui doive la moindre ligne. Antonio 'der Wale' était le dernier représentant à Cracovie et à Breslau d'une époque révolue: le document conserve le souvenir de sa puissance matérielle et de son influence politique ('par la grâce de Dieu, je possède suffisamment de châteaux et de villages'), et enregistre à sa manière l'incrédulité des contemporains devant sa chute ('j'ai conservé de grandes richesses cachées, de quoi faire vivre des centaines de personnes').[80]

CONCLUSION

Il est salutaire de débarrasser l'histoire des légendes qui l'envahissent et risquent d'obscurcir la compréhension des phénomènes. Mais l'histoire se construit à chaque génération avec des matériaux anciens et nouveaux; elle ne doit faire fi d'aucune source; et de même que le prospecteur, l'oreille au sol, repère les cours d'eau souterrains, de même on retrouve, à travers des textes aussi composites que les 'Walenbücher', la démarche et les observations des hommes de terrain à la fin du Moyen Age.

Dans son rapport sur les mines vosgiennes publié par Sebastian Münster en 1550, Johannes Haubensack se demandait pourquoi l'histoire avait fait si peu de place à l'exploitation minière, alors que le sujet est merveilleux ('wunderbares Ding') et que de grandes entreprises se sont développées en tous lieux d'Europe. Il donne lui-même la réponse quelques lignes plus loin: 'Je vois la raison principale dans le fait que personne n'est en mesure d'écrire vraiment sur un sujet qu'il n'a pas lui-même vu ou éprouvé.'[81] Conscient des difficultés de la description technique—qu'il laisse à d'autres—,[82] Haubensack décrit, gravures à l'appui, le travail des hommes qu'il connaît bien. Il est bien vrai que l'histoire s'écrit dans une chambre, comme la philosophie, dans un poële. Mais qu'il s'agisse, comme l'écrit Münster, des 'mystères de la nature', et les documents valent la promenade.

Sans doute, ne saisit-on à travers les 'Walenbücher' qu'un reflet terni des curiosités, des espérances et des efforts physiques que les prospecteurs les plus avertis partageaient avec tous les chercheurs de 'trésors'; quant aux connaissances pratiques que mineurs et métallurgistes se transmettaient oralement ou parfois par écrit, ces textes en perpétuent l'écho personnel; un siècle plus tard, l'héritage accumulé fut systématiquement présenté dans les grands traités d'érudition technique. Demeure le témoignage d'une quête, et l'évocation de paysages d'altitude, avec leurs alpages, leurs torrents, les sentiers qui conduisent de roches en cîmes, et les traces solitaires d'anciens travaux miniers.

Philippe Braunstein
EHESS, Paris

ANNEXES

Les textes présentés et traduits ont été publiés à plusieurs reprises. Le texte (I) est un guide topographique souterrain explorant le Weingartenloch ('Trou des vignes'), près de Bartholfelde, dans le Harz méridional. Datant de la fin du XVIIe siècle, il a été publié d'après un manuscrit par F.E. Bruckmann, *Magnalia Dei in locis subterraneis* I (Brunswick, 1727), et reproduit par G. Heilfurth, *Bergbau und Bergmann in der deutschsprachigen Sagenüberlieferung, I, Quellen* (Marburg 1967), dans la section consacrée aux 'Walenbücher', n° 987; il a été à nouveau publié par H. Wilsdorf et R. Schramm, *Venetianersagen* (VEB, Leipzig, 1985), 86-8, en compagnie de deux récits légendaires relatifs au même site, une montagne de gypse trouée de cavernes et qui servit de carrière pour la construction du monastère de Walkenried. Les textes (II) et (III) sont des itinéraires de prospecteurs, extraits des 'Walenbücher' silésiens. Le texte (II) relatif à la région de Hirschberg, a été publié par K. Schneider, Die Walen im Riesengebirge, *Mitteilungen des Vereins für Geschichte der Deutschen in Böhmen*, LX, Prague, 1922, HH V, 309-12. On a indiqué plus haut le contexte; il aurait été rédigé après 1650, mais son contenu date des années 1460. Le texte (III) relatif à la zone minière de Silberberg-Reichenstein a été publié par K. Wutke, *Codex diplomaticus Silesiae, XX, Schlesiens Bergbau und Hüttenwesen 1136-1528* (Breslau, 1900), 'Wegweiser zu den Bergwerken in der Oberlausitz und in Schlesien,' 85-7.

TEXTE I

Pour y entrer, prends tout de suite à gauche, tu trouveras ce signe (V),[1] c'est là qu'il faut descendre profondément, puis fais une douzaine de pas, entre en rampant à droite, et tu vas trouver la descenderie. Là, tu peux te tenir debout sur une roche émergée de deux doigts, car il y a là un cours d'eau. Remonte-le en rampant, mais si tu ne peux pas passer, reviens vers la roche et aussitôt vers la gauche, tu vas entrer dans une étroite galerie; poursuis-la jusqu'à ce que tu rencontres des roches grises, sur lesquelles tu verras un (V) en hauteur. C'est là que juste devant toi une châtière te permet de descendre; passe-la, et quand tu seras en bas, tout de suite à gauche engage-toi en rampant sur le ventre (le passage n'est pas plus long que quelques maisons) et tu buteras sur un mur; en t'y appuyant,[2] tu trouveras un (V) qui t'indiqueras de remonter; continue, et tu arriveras dans un étroit conduit, qui se terminera par un trou bouché avec des pierres; tu les déplaces, et quand tu pourras passer, tu t'y glisses, et arrives dans une galerie d'environ 60 m. Là, tu rencontreras deux mineurs portant des lampes,[3] passe hardiment devant eux sans parler; ils te cèderont le passage, et si tu continues, tu arriveras dans des roches blanches. Là, tu trouveras un trou rond, que tu traverses en rampant, puis cela s'élargit, tu continues jusqu'à un moine, qui se tient à un coin, un pic à la main, t'indiquant la présence de l'eau. Quand tu arrives près de l'eau-près de deux mêtres de largeur-il y aura deux planches en travers, passe-les, tu verras à gauche une roche noire, qui renferme de l'argent natif, et si tu en détaches un morceau, il brillera vivement; et si tu le soumets au feu de la lampe jusqu'à ce qu'il noircisse, tu entendras une petite détonation . . .

1. une main schématique indiquant une direction
2. à contre-pied (?)
3. probablement, comme le moine plus loin, des figures gravées.

TEXTE II

Quand tu arrives près de Hirschberg,[1] demande le chemin de Warmbad, où tu pourras te baigner, afin de ne pas éveiller de soupçons, car les gens de la région du Kynast attachent beaucoup d'importance à la chose. En te baignant, pose des questions sur un village qui s'appelle Seifershau, et ne te dévoile pas; car ce village est habité par des charbonniers qui brûlent la cendre, pour faire du verre. Ils ont peu de complaisance pour les gens, et peu de relations avec leurs semblables. Il y a une église dans le village, et sur le flanc de l'église se dresse un moulin, près duquel tu verras deux chemins: celui de droite va à Geiritz, ne l'emprunte pas; prends celui de gauche, qui va vers la Combe des Anges; et si quelqu'un devait te questionner, réponds que tu es un cueilleur de simples; car ils savent bien que poussent par là bien des espèces de plantes et de racines. Quand tu seras arrivé dans la combe, va droit sur le grand chemin; car il y a aussi un petit chemin, qui contourne la montagne noire que tu vois étendue devant toi; c'est celui qu'empruntent la plupart des gens qui cherchent des racines. Tu ne le prends pas, mais au contraire, le grand chemin. Et quand tu es monté, alors tu vois se dresser devant toi une grande montagne; dirige toi vers elle, et tu verras devant toi une croix de pierre. Quand tu arrives devant la croix, dirige toi vers la droite, à une bonne portée d'arquebuse, puis à gauche, et tu arrives dans une prairie, dans laquelle il y a beaucoup de chemins et bien des possibilités; je n'ai rien appris d'inquiétant sur ce point. Mais tu dois te tenir à main droite pour arriver à un endroit verdoyant, qui ressemble à une verte prairie. De là, tu arrives au bord d'un cours d'eau, que tu remontes pendant un court moment, jusqu'au point où tu verras que l'eau se divise, et tu remonteras le plus petit ruisseau, et à deux jets de pierre, tu arriveras à une grosse roche. Sur cette roche sont gravées plusieurs figures, en forme d'un homme, d'un chien, d'un écu, d'une croix et d'autres signes. Ne te laisse pas détourner par ces signes, mais va de la roche vers le midi, à un jet de pierre, vers un autre roche.

Escalade-la, et regarde vers l'Est, tu verras une roche qui a apparence humaine; dirige-toi vers elle vers l'Est, tourne tes regards vers le midi; or cette roche présente deux pointes, comme une fourche, vers le couchant et vers le midi, elle est creuse par-dessous. Les signes ne sont là que pour égarer. Tu ne peux vraiment questionner personne, mais la roche se dresse dans une prairie, et si tu fais un pas ou deux à partir de la roche, allonge-toi l'oreille contre la terre, et tu entendras une chute d'eau, comme si cela tintait d'une pierre à l'autre. Alors creuse ici ou là, dans la direction du gîte, tu trouveras des grains d'or, gros comme des pois ou plus petits. Mais il y a bien des exploitations sur d'autres chemins, de sorte que je crois que personne ne prête d'attention à ce gîte, et je ne l'aurais pas non plus trouvé sans quelques indications que je portais avec moi. Je reconnais en conscience que j'ai trouvé quelques morceaux d'or natif, et gros comme des noix d'Italie, et même assez nombreux, et si l'on m'objectais que tant d'années se sont passé, et que tout a dû être emporté, en toute vérité, je dis qu'il n'en est rien. Plus il y a d'eau, plus de l'or s'accumule; c'est une chose que j'ai vérifié. Car mon grand-père qui a vécu à Florence la grande m'a révélé ce qu'il n'avait pas révélé à son fils, mon père, et il m'a raconté qu'il avait lavé l'or dans des bassins, et je l'ai moi-même vérifié, en l'an 1656 (*sic*),[2] année où moi-même j'ai tant gagné qu'avec les miens je ne désire plus rien pour mon entretien et ne veux plus mettre ma vie en danger. *Item.* Plus les eaux sont grosses, plus tu trouves d'or, car cela vient de gîtes précieux, et c'est emporté des hautes montagnes par l'écoulement de l'eau. *Item.* Dans la direction de la fourche, il y a aussi de bons gîtes, que tu pourrais trouver; il y a dans la région plus d'or que ne vaut l'ensemble de la Silésie; mais c'est demeuré inconnu aux gens du pays, et personne ne le recherche; je n'entre pas dans les raisons, ne serait-ce que parce qu'il n'appartient pas à n'importe qui d'en tirer profit.

Pour ma part, avec l'aide de Dieu, j'ai fait en mon temps toutes ces recherches sans être mis en péril.

Plus loin vers le midi, se dresse une montagne qui s'appelle le Riesenberg. Si tu veux prendre cette direction, tu trouveras une grande roche; devant cette roche, tu en verras une autre, près de laquelle passe un sentier, qui va dans la Combe d'Aupa:[3] il est dangereux de s'y rendre, car bien des prospecteurs y sont morts de faim, faute d'avoir pu retrouver leur chemin. J'y ai été moi-même, et y ai vu bien des ossements; près de certains d'entre eux, j'ai trouvé des pierres précieuses de prix. Si tu voulais t'y rendre, fais comme j'ai fait moi-même; prends avec toi de bons et fidèles compagnons, qui ne soient pas des mauvaises têtes, mais qui s'acceptent mutuellement, avec fidélité et sans équivoque; emportez du pain et des sucreries autant que vous pouvez et au moins pour huit jours; l'eau ne vous manquera pas sur place, et bonne à boire. Car vous ne rencontrerez pas des êtres réels, ce sont des compagnons repoussants, dont les humains ne peuvent se rapprocher. Alors entre dans ce secteur avec tes compagnons en invoquant le nom de Dieu, et quand tu arriveras dans la combe, alors prends à main gauche, et tu te trouveras près d'un petit gîte, à peine quatre aunes de largeur, puis près de l'eau, fais à peu près une lieue de chemin, tu arriveras à une croix, d'où partent plusieurs chemins. Continue à suivre l'eau jusqu'à ce que tu atteignes une grosse roche. Quand tu arrives près de la roche, traverse le cours d'eau et rejoins le grand chemin, sur l'autre rive; tu arriveras à une roche qui a trois marches; près d'elle, commence à monter et suis le sentier jusqu'à ce que tu arrives à un autre cours d'eau qui s'appelle le Zacken; remonte-le, jusqu'à ce que arrives à une grande roche, pour éviter les chemins peu sûrs en raison de leur nombre, qui risquent de t'égarer; c'est pourquoi je ne peux te donner d'instructions plus précises que de suivre le cours d'eau. Quand tu seras arrivé à la grande roche, prends le prochain sentier près de l'eau; tu verras un endroit vert; fouille à travers la mousse, et tu trouveras de l'or natif. Au même endroit, s'étend une montagne, que l'on peut escalader par degrés; quand tu arriveras au sommet, tu trouveras un étang.[4] Mais si tu veux y monter, vois à ne t'y rendre que par temps clair, car s'il pleut, tu es perdu avec tous tes compagnons; si tu remarques que le temps se met à la pluie, pendant que tu es sur le chemin, rebrousse chemin, je te le conseille. *N.B.* Près de la roche qui est baignée par le Zacken, cherche à main gauche, tu trouveras un petit gîte, tu y trouveras de belles pierres, comme des améthystes, des hyacinthes, des émeraudes, des chalcédoines et autres pierres précieuses. Mais cherche toujours, où que tu te rendes, à bien noter chemins et sentiers à l'aller, afin que tu les trouves au retour; fais ce que tu peux pour ne pas t'égarer; mais si tu te trompes sur le chemin du retour, tu auras bien du mal à revenir parmi les hommes.

Item. Ne te préoccupe pas, si des chimères d'esprits et de fantômes surgissent devant toi, il ne peut rien t'arriver.

Cette lettre-itinéraire, Nicolas Orler l'a obtenue lorsqu'en 1650 il avait été envoyé par son maître pour ses affaires dans le bas pays, de quelques paysans, pour une somme dérisoire, parce qu'ils ne connaissaient rien à la chose, et qui avaient conservé non loin de la mer ce petit livre, qu'ils avaient trouvé et apporté avec eux.

1. L'actuel Jelenia Góra. L'itinéraire a été parfaitement élucidé à partir de Hirschberg. Plusieurs légendes silésiennes se déroulent dans les lieux réels évoqués par l'itinéraire.

2. Le manuscrit de Trautenau, qui conserve une autre copie du même texte, porte la date de 1466. K. Schneider (note 15), p. 291.

3. La première grande roche est le Koppenplan (1430 m), la seconde, la Kleine Sturmhaube (1436 m); la combe d'Aupa (Appe), à 1200 m, a été un lieu de production de vitriol au 16° siècle, d'arsénic vers 1920; le texte la décrit comme déserte, ce qui suffit à corriger la datation (note 2 ci-dessus).

4. Il s'agit du Hohe Rad (1506 m).

TEXTE III

A partir de Schweidnitz, demande le chemin de Reichenbach,[1] ensuite longe la montagne à main droite jusqu'à un village, qui s'appelle Langbielau, puis dirige toi vers Lampersdorff, puis vers Schönwald. Là, va jusqu'au bout du village, tu trouveras un chemin pour monter au Silberberg. Là, le long de la route, tu trouveras du plomb natif, parfois comme des fèves, parfois comme des pois, plus ou moins gros. Et sur la montagne près de la route, à main gauche, se trouve une pente argileuse, et dans l'argile, tu trouveras aussi du plomb, comme ci-dessus. Si tu veux chercher des pierres, reviens en arrière, retraverse Schönwald et tu arriveras rapidement à Frankenstein. Là, dirige-toi vers la porte de Breslau, et demande le Kummerberg, il est à trois quarts de lieue de Frankenstein. Tu trouveras là toutes sortes de pierres précieuses. Mais reprends la direction de Frankenstein et demande le chemin de Reichenstein. Là, tu trouveras des scories d'or et des scories d'argent, par milliers des cuveaux. Si tu peux en tirer parti, grâce à ton savoir-faire, fais-le. Et l'on pourra te renseigner sur les travaux que l'on a fait là avant la guerre hussite, avec le plomb décrit plus haut, et avec le minerai qu'on trouve sur place. Le minerai de Reichenstein ne vaut rien sans le plomb, et on a fait beaucoup d'essais, auxquels j'ai assisté. On n'arrivait à rien quand se présentait du 'cobalt' blanc, dont j'ai eu plusieurs échantillons entre les mains. C'est pourquoi il faut utiliser le plomb que j'ai décrit plus haut, comme me l'a dit un vieux mineur de Schönwald.

Mais si tu veux aller prospecter en haute montagne, demande la direction pour aller de Reichenstein à Friedberg. Il n'y a là qu'un seul chemin, qui fait trois lieues par Goldenstein. Quand tu auras fait les trois quart du chemin de Friedberg, tu seras devant deux anciennes verreries. Fais encore un quart de chemin, et regarde autour de toi des deux côtés, et tu trouveras une plante, qu'on appelle la Vulve de chienne. Cette plante a l'allure d'une tête humaine, avec son visage, et des feuilles comme celles des chardons, mais plus vertes et s'élevant plus haut que celles enroulées des chardons. A quoi peuvent servir ces plantes, tu pourras l'apprendre. Puis monte sur le Bogenberg, quand tu arrives sur la route qui va de Freiwaldau sur le Goldenstein, traverse la route et dirige-toi vers les hauteurs que tu vois, et qui s'appellent le Kalenberg. Là tu trouveras un sentier sur la montagne; suis-le, jusqu'à ce que tu le perdes. Ensuite suis plus ou moins la direction du Sud, jusqu'à ce que tu vois devant toi une roche, puis une autre. Descends la pente devant toi vers la droite, entre un et trois arpents, et tu arrives sur un gîte qui plonge vers le sud. Tu y trouveras des pierres noires, allongées et anguleuses; et si tu les casses, elles sont à l'intérieur brunes comme de l'écarlate, et elles sont lourdes et dures. Et tu peux expérimenter ce qu'elles peuvent contenir. Puis retourne sur tes pas, jusqu'à ce que tu reviennes à la route et emprunte-la pour aller à Freiwaldau. De Freiwaldau vers le Spitzenstein, il y a un mille. Si de là tu veux aller dans la montagne tu auras des chances de trouver de l'argent et de l'or. Si tu veux le tenter, la montagne est à l'intérieur pleine de ressources. Si tu veux l'explorer, c'est à toi de t'aventurer, puisque, comme me l'a dit Procopius Hoberg qui a été dans la montagne, il y a beaucoup de galeries à l'intérieur. Et dans la montagne, il y a un *filon*,[2] et si quelqu'un veut s'y risquer et y travailler, il trouverait ce que j'ai dit. Et celui qui attaquerait le filon par dessus à l'intérieur de la montagne, et qui voudrait le tenter, il trouverait de quoi devenir un puissant personnage, si Dieu le voulait. Il m'a aussi dit qu'un docteur avait tenté l'aventure et avait acquis de grands trésors.

Depuis le Spitzenstein, tu as encore un quart de chemin à faire vers Saubsdorf; là, demande le chemin de l'Ylmenberg qui se trouve près de Rothwasser, là tu trouveras un puits carré que les bourgeois de Neisse ont exploité, puis ils ne se sont pas entendus, et ont laissé les choses en plan. J'y ai vu à l'intérieur de l'argent natif, les fions s'étiraient, de la largeur d'un dos de couteau.[3] Celui qui voudrait travailler là aurait l'accord du seigneur de Kaltenstein, à un quart de chemin de là. Dirige-toi vers Weidenau, et sors de la montagne où tu voudras, et rends grâces au Dieu tout-puissant. C'est Procopius qui m'a donné le livre, et qui l'avait utilisé pour ses recherches.

1. Tous les noms de lieux cités sont identifiés.

2. L'interprétation paraît plus satisfaisante que celle de l'éditeur, qui traduit 'flis' par 'torrent'; il semble qu'on peut entendre 'flöz'.

3. 'Eyn gut messirrucke': l'interprétation littérale est sans doute la plus satisfaisante.

Notes et Références

1. Le mot de paysage n'apparaît dans la langue française que vers 1550. Pour la ville, cf. P. Lavedan, *Représentation des villes dans l'art du Moyen Age*, 1954, cité par B. Chevalier, 'Le paysage urbain à la fin du Moyen Age: imaginations et réalités', *Le paysage urbain au Moyen Age, Actes du XIe Congrès des historiens médiévistes de l'enseignement supérieur* (PUL, Lyon, 1981), 7–21. Pour une position du problème de la description, cf. Ph. Braunstein, 'Le Sinaï décoloré. Vision, langage et paysage à la fin du Moyen Age', *De l'Autre, Cahiers de psychologie de l'art et de la culture*, 10, 1984, 25–32.

2. Sur les routiers, cf. Ch. Schefer, *Recueil de voyages et de documents pour servir à l'histoire de la géographie depuis le XIIIe jusqu'à la fin du XVIe siècle* (Paris, 1892); sur la plus ancienne carte routière en pays d'Empire, H. Krüger, *Des Nürnberger Meisters Erhard Etzlaub älteste Strassenkarte von Deutschland, Jahrbuch für fränkische Landesforschung*, 18, 1958. Un exemple d'itinéraire à des fins de politique commerciale, le réseau des routes fluviales pour l'exportation du sel de Transylvanie au XVIᵉ siècle: J. Strieder, 'Ein Bericht des Fuggerschen Faktors Hans Dernschwam über den Siebenbürgener Salzbergbau um 1528', *Ungarische Jahrbücher*, 13, 1933, 259–90.

3. Cf. G. Fournier, *Châteaux, villages et villes d'Auvergne au XVe siècle d'après l'armorial de Guillaume de Revel* (Paris, 1973); J. Mossay, 'L'Avesnois et les albums de Croy', *Mémoires de la Société archéologique et historique d'Avesnes*, 1962, p. 11 et suiv., cité par G. Sivéry, La description du paysage rural par les scribes et les paysans du Hainaut à la fin du Moyen Age, *Le paysage rural: réalités et représentations, Actes du Xe Congrès des historiens médiévistes de l'enseignement supérieur public, Revue du Nord*, LXII, 1980, 61–71. C'est un conflit sur l'usage de la voie d'eau (flottage du bois et (ou) usines hydrauliques) qui est à l'origine du dessin aquarellé représentant le cours de la Mulde en Saxe: H. Wilsdorf, W. Herrmann et K. Löffler, *Bergbau. Wald. Flösse, Freiberger Forschungshefte D 28* (Berlin, 1960).

4. Outre P. Lavedan, *Histoire de l'urbanisme* (Paris, 1959), cf. J. Schulz, 'Jacopo

de' Barbari's View of Venice: Map Making, City Views and Moralized Geography before the Year 1500', *The Art Bulletin*, LX, 1978, 425–74.

5. Pour la référence antique, M. Cantor, *Die römischen Agrimensores* (Leipzig, 1878). Sur les mesures de parcelles, les spécialistes de l'arpentage, les efforts d'unification d'unités de surface dans les campagnes du Hainaut entre le XIIIe et XVIe siècle, cf. G. Sivéry, *Structures agraires et vie rurale dans le Hainaut à la fin du Moyen Age* (Lille, 1977), 1: 58 et suiv. Les plus anciens documents cartographiques du monde germanique sont liés à l'exploitation du sel près de Hallein (Salzbourg): cf. F. Kirnbauer, 'Die ältesten Dokumente des Markscheidewesens', *Montanistische Rundschau*, 27, 1935, n° 20, et R. Palme, 'Die Weiterentwicklung des österreichischen Grubenvermessungswesens im Spätmittelalter und zur Beginn der Neuzeit', *Ingenieurvermessung von der Antike bis zur Neuzeit, 3. Symposion zur Vermessungsgeschichte* (Dortmund, 1987), éd. par H. Junius, Stuttgart, 1987, 141–156, 141. Que des conditions nouvelles d'exploitation minière entraînent des formes juridiques nouvelles, qui se traduisent dans l'arpentage, est l'une des conclusions de M. Ziegenbalk, 'Aspekte des Markscheidewesens mit besonderer Berücksichtigung der Zeit von 1200 bis 1500', *Montanwirtschaft Mitteleuropas vom 12. bis 17. Jahrhundert: Forschungsprobleme, Der Anschnitt, Beiheft 2*, éd. par W. Kroker et E. Westermann (Bochum, 1984), 40–9.

6. La projection des galeries sur la surface du sol est désignée dans la région de Salzbourg par le terme usuel en langue vernaculaire de 'Tagschaftricht'; cf. *Salzburger Urkundenbuch* 4, 56: *debita diei linea, que vulgo tagschaftriht dicitur, ut ipsi perspectis montibus et fodinis sub terra et mensuratis super terram litem deciderent* (1266) . . ., cité par H. Klein, 'Zur Geschichte der Technik des alpinen Salzbergbaues im Mittelalter', *Mitteilungen der Gesellschaft für Salzburger Landeskunde* 101, 1961, 261–8.

7. La précision est de l'ordre de la largeur d'une maison: *Ibid.*, 268 (1270). Les écarts mesurés entre une coupe orthogonale d'un site du Harz en 1722 et aujourd'hui s'expriment en centimètres; cf. Ziegenbalk, *op. cit.* (5), 48, qui attire l'attention sur les variations d'échelle que l'on relève dans de superbes documents de synthèse, comme la représentation des mines du Harz par Zacharias Koch en 1606, comparables dans leur esprit à des vues cavalières. Daniel Flach a tenté, par un dessin aquarellé de 10 m de long, de représenter les 4 km qui séparent Wildemann et Zellerfeld dans le Harz, à la fois en coupe et en perspective, en surface et en profondeur, la virtuosité du dessin ne pouvant prétendre à la rigueur d'un relevé topographique: *Der grosse Grubenriss des Oberbergmeisters und Markscheiders Daniel Flach von den Gruben und Wasserlösungsstollen auf dem Zellenfelder Hauptgange zwischen den Bergstädten Wildemann und Zellerfeld anno 1661* (Westfalia, Lünen, 1974). Sur les représentations du milieu souterrain dans l'espace vosgien, signalons des recherches en cours de J. Grandemange et de B. Ancel; cf. B. Ancel et P. Fluck, 'Archéologie souterraine et spéléologie minière', *Bulletin de la Société industrielle de Mulhouse*, 1987, 123–8.

8. G. Agricola, *Zwölf Bücher vom Berg- und Hüttenwesen* (DTV, Munich, 1977), livre II, 26 et suiv.

9. Ces prescriptions font partie des lieux communs à la fin du XVI^c siècle; elles sont résumées en quelques lignes par Wilhelm Prechter, entrepreneur minier dans le Val de Lièpvre vosgien, dans le rapport qu'il rédigea en 1602 pour son maître, le sire de Ribeaupierre: H. Winckelmann, *Das Bergbuch des Lebertals* (Westfalia, Lünen, 1956), 61, 141.

10. Un exemple emprunté à un secteur minier sur lequel nous allons revenir: l'évêque de Breslau concède à Valt Haugwolt de Zugmantel 'ein alde vorfallene grube uffim Czugmantel': *Schlesiens Bergbau und Hüttenwesen (1136–1528)*, éd. par

K. Wutke, *Codex diplomaticus Silesiae, XX* (Breslau, 1900), n° 317 (1506/5/X). Dans le secteur vosgien, 'ce qui a déterminé les mineurs à effecteur des prospections, ce sont les anciens puits', déclare Johannes Haubensack dans son *Historia und Cronning*; passant en revue les travaux récents, il raconte comment la fosse abandonnée du vieux Saint-Guillaume fut reprise par quelques mineurs 'qui avaient travaillé dans la concession de messire Reinhart et, ainsi, connaissaient parfaitement les lieux'; avant que le concessionnaire s'en soit rendu compte, ils avaient atteint en grande profondeur des gîtes très riches, 'comme un détachement armé assaille l'ennemi dans son propre camp'; cf. Winckelmann, *op. cit.* (9), 69, 75.

11. Sebastian Münster, *Cosmographia, Das ist Beschreibung der gantzen Welt . . .*, (Bâle, 1628; fac-similé, Lindau, 1984), I, 808.

12. Cf. G. Heilfurth, *Bergbau und Bergmann in der deutschsprachigen Sagenüberlieferung Mitteleuropas* (Marburg, 1967), en particulier la substantielle introduction (47-196) au répertoire raisonné et aux *indices* à plusieurs entrées. Hors du domaine germanique, demeure important P. Sébillot, *Les travaux publics et les mines dans les traditions et les superstitions de tous les pays* (Paris, 1894). On peut évoquer dans ce contexte les discussions savantes autour de la figure de Rübezahl, incarnation silésienne de Wotan, démon de la fécondité souterraine, dont la carrière et les attributions sont étroitement liées à l'exploitation minière; cf. A. Lincke, *Die neuesten Rübezahlforschungen: Ein Blick in die Werkstatt der mythologischen Wissenschaft* (Dresde, 1896), et P. Regell, 'Wanderungen und Wandelungen der Rübezahlsage', *Mitteilungen der schlesischen Gesellschaft für Volkskunde*, 18, 1916, 165-226.

13. Agricola s'est exprimé à plusieurs reprises sur la présence réelle des esprits dans les mines, en particulier dans son traité de 1549, *De animantibus subterraneis* (édn. allemande, DTV, Munich, 1977), et les décrit en deux groupes, les uns, effrayants et pernicieux, les autres, bienveillants et propices (542-3), sans aucune référence diabolique.

14. Heilfurth, *op. cit.* (12), Index 'Venediger', n°s 142-5.

15. Outre la section qui leur est réservée dans le volume de Heilfurth, *op. cit.* (12), Der Reichtum der Gebirge und Gewässer. I, Wegweisung zu Fundplätzen ('Walenbücher') 842-63, ces recueils topographiques sont l'objet d'une annexe au volume *Venetianersagen. Von geheimnisvollen Schatzsuchern*, éd. par R. Schramm et H. Wilsdorf (VEB, Leipzig, 1985), études, extraits et bibliographie, 217-81. Les 'Walenbücher' silésiens ont été partiellement publiés par K. Wutke, *Codex diplomaticus Silesiae, XX*, . . . 'Wegweiser zu den Bergwerken in der Oberlausitz und in Schlesien', 83-7, et par K. Schneider, 'Die Walen im Riesengebirge', *Mitteilungen des Vereins für Geschichte der Deutschen in Böhmen*, LX, Prague, 1922, 276-314, part. 302-14.

16. Schneider, *op. cit.* (15), 286; W.E. Peuckert, 'Walen und Venediger', *Mitteilungen der schlesischen Gesellschaft für Volkskunde*, 5, 1929, 205-47, part., 216.

17. Wilsdorf, *op. cit.* (15), 219 et suiv., rappelle que le manganèse n'entre dans la classification des métaux qu'en 1774 et que le cobalt a passé pendant des siècles pour un minerai d'argent prometteur et décevant, avant d'être isolé en 1735 (qualités et défauts qui lui ont valu son nom d'esprit malicieux): le manganèse contribuait à la transparence du cristal, le cobalt, au bleu profond. En l'absence de sources relatives au bleu des vitraux français du XIIᵉ siècle, il paraît tout-à-fait aventureux de se fonder sur les écrits d'Antonio da Pisa et de Cennino Cennini pour interpréter l'‘azzuro d'Alamagna' comme un bleu de cobalt, dont l'unique provenance en Occident serait, au 14ᵉ siècle comme deux siècles plus tôt, les Monts Métallifères.

18. Peuckert, *op. cit.* (16), 244, montre parfaitement que les Vénitiens

n'apparaissent pas dans les plus anciens 'Walenbücher': si le nom et les légendes qui l'entourent avaient été connus en Silésie avant le XVIe siècle, on n'aurait pas appelé 'Walen' (Welsches) les prospecteurs itinérants.

19. Chez G. Meyer, *Bergwercks Geschöpff und wunderbare Eigenschafft der Metalsfrüchte* (Leipzig, 1595), 42, l'intéressante contamination de l'image des prospecteurs de minerais avec celle des 'fahrende Schüller' et autres 'Landfahrer' (routiers?). Dans son ouvrage, *Gründliche Beschreibung des Fichtelgebirges*, publié en 1542, Caspar Bruschius explique que des peuples très éloignés, 'comme les Welsches, les Tziganes et les Espagnols connaissent parfaitement les montagnes allemandes, en particulier le Fichtelgebirge, où l'on a trouvé nombre d'ouvrages écrits en welsche, en français et en allemand des Pays Bas, décrivant et désignant les lieux où l'on trouve l'or, les perles et les pierres précieuses'. L'allusion aux Tziganes serait à mettre en relation avec l'orpaillage en Transylvanie: Wilsdorf, *op. cit.* (15), 232.

20. Le plus catégorique, cité par Wilsdorf, *op. cit.* (15), 224, est le maître général des mines du royaume de Bohême, Lazarus Ercker von Schreckenfels, qui publia à Prague en 1574 *Beschreibung aller fürnehmisten Erzt und Bergwercksarten*. Il affirme que les étrangers ne viennent en terre d'Empire que pour en extraire des minerais bruns ou noirs, vitrifiés quand on les casse, et dont on n'extrait pas d'or, mais des 'additifs' ('Zusatz') pour en faire des colorants ou du verre.

21. Lazarus Ercker (note précédente) n'avait personnellement jamais rencontré de prospecteurs étrangers. Agricola, vingt ans plus tôt, mentionnait comme une évidence la présence d'Italiens dans les massifs montagneux allemands, lavant le sable des ruisseaux pour en extraire les paillettes d'or et les grenats: *op. cit.* (8), VIII, 291.

22. Lorsque l'abbé de Heinrichau accorde en 1454 près du Silberberg une concession, 'is sey off golt silber erz adir edilgesteine', il précise que ce sera pour 'sichern, wasschen, suchen und bauen' (cribler, laver, prospecter et foncer), toutes formes d'exploitation que des chercheurs d'or pouvaient mettre en oeuvre sans mises de fonds importantes: Wutke, *op. cit.* (10), n° 195. En Silésie comme en Bohême, la fouille 'sauvage' était interdite: cf. A. Steinbeck, *Geschichte des schlesischen Bergbaues, seiner Verfassung, seines Betriebes*, 2 vols (Breslau, 1857), I, 79 (règlement minier de Löwenberg, 1278). En Saxe, le capitaine de Zwickau fait arrêter en 1564 trois individus, dont un Bohémien et un Styrien, qui font de la prospection sauvage et se vantent d'avoir un 'Walbüchlein', dont ils suivent les indications; cf. H. Schurtz, 'Der Seifenbergbau im Erzgebirge und die Walensagen', *Forschungen zur deutschen Landes- und Volkskunde*, 5, 1890, 129.

23. Sur ce thème, cf. Wilsdorf, *op. cit. (15)*, 68–70, 199 et suiv.; Heilfurth, *op. cit.* (12) 684, 756, 765–70, 776, 782, 785, 793, 802, 871, 874, 888, 954, 960; le miroir permet de voir que les minerais ne sont pas 'mürs', 907.

24. Bergers, charbonniers, bûcherons occupent innocemment des lieux déserts, chargés d'histoire minière; cf. Wilsdorf, *op. cit.* (15), 102–4, 114.

25. R. Altmüller et F. Kirnbauer, *Ein steirisches Walenbüchlein, Leobener Grüne Hefte*, 125 (Vienne, 1971).

26. Schneider, *op. cit.* (15), 280, signale que le musée de Hirschberg (Jelenia Góra) possédait une série de blocs de pierre comportant dessins et signes; les inscriptions dites 'Walenzeichen' de Silésie sont reproduites par Heilfurth, *op. cit.* (12), 161, et par Wilsdorf, *op. cit.* (15), 275–8.

27. Fink, 'Die Bergwerksunternehmungen der Fugger in Schlesien', *Zeitschrift für Geschichte und Altertum Schlesiens*, 28: 312. et suiv.

28. C. Faulhaber, *Die ehemalige schlesische Goldproduktion mit besonderer Berücksichtigung des Reichensteiner Bergreviers* (Breslau, 1896), 23–9.

29. L'ouvrage de Steinbeck, *op. cit.* (22), n'a pas été remplacé; les 'monuments' du droit minier sont chronologiquement séparés par des plages d'ombre. L'analyse des documents publiés par Wutke, *op. cit.* (10), atteste la pauvreté de nos connaissances sur telle ou telle mine de première importance, par exemple Goldberg: le duc de Silésie fait contrat avec le prêtre Michel de Deutsch Brod, curé à Prague, pour la remise en état des installations ennoyées, en 1404. On ne connaît ensuite que deux actes de concession, en 1420 et en 1477. Reichenstein et Zuckmantel n'ont véritablement d'histoire qu'à partir des années 1480.

30. Peuckert, *op. cit.* (16).

31. Schneider, *op. cit.* (15).

32. J. Schickfuss, *Neue vermehrte schlesische Chronika*, IV, 3, p. 12 et suiv.

33. J. Praetorius, *Satyrus Etymologicus* (préface); cf. K. Zacher, 'Rübezahl-Annalen bis Ende des 17. Jahrhunderts', *Festschrift zur Feier des 25 jährigen Bestehens der Ortsgruppe Breslau*, 1906, 96.

34. *Beschreibung deren Wallensern welche in Teutschland Gold, Silber Ertz gesucht und zu Nutzen gemacht.*

35. Schneider, *op. cit.* (15), 304, HH III, 'in dem selbigen Schachte findest du gediegene Goldstücke als die Hünereyer gross; ist wahrhaftig wahr'.

36. *Ibid.*, 300, HH V, 'Die zeichen sind gemacht um der irrung willen'.

37. *Ibid.*, 308, HH IV, 'Dis hab ich obgemeldeter Hans Mann von Regenspurg zweymal gefunden aber übel angewendet, derhalben mich Gott gestrafft hat, das ichs zum 3ten mahl nicht finden können.'

38. Wutke, *op. cit.* (10), 83, 'Ich Anthonius Wale vormelde gote zcu lobe manchem armen zcu trosste und meyner zele zcu seligkeyt weme seyn muc hercze und begyr stehet noch gutte und noch ere der froge nach einer stad . . .'.

39. *Ibid.*, 'wer do hen gehen wil adir suchen der sal sich alzo dorczu bereytin daz her alle recht tuhen sal gleycher weyse alzo eyner sterbin sulde und sal ij tage fastin zcu wasser und zcu brothe . . .'.

40. Schneider, *op. cit.* (15), 302, BH I, 'so vorsuche dy steyne mit eyner keylhowe zo sint etzliche bemost ynnewingk sint sy pur golt'; 303, HH III, 'darinnen findest du gutt gewachsen gold, ist arabisch gold, ist ganz schwartz und wächst allda . . .'; G. Heilfurth, *op. cit.* (12), p. 844 (Weingartenloch/Harz), 'ein schwartzer Fels der gemeiniglich gediegen Silber hält; und wenn du davon etwas loss machest, so wird es helle gläntzen, machst du es aber mit dem lichte schwartz, so wirds einen Schall von sich geben . . .'.

41. Heilfurth, *op. cit.* (12), N° 824, 828, 831, 852, 865, 908, 929, 938, 957; Wilsdorf, *op. cit.* (15), N° 7, 56, 61.

42. Wutke, *op. cit.* (10), 83.

43. Les conseils relatifs au calendrier (Schneider, *op. cit.* (15), 303, BH I et HH III) font partie du répertoire des légendes; cf. Heilfurth, *op. cit.* (12), Index des thèmes, 1026.

44. Schneider, *op. cit.* (15), 308 (HH IV).

45. Sous des apparences diverses, figure de moine gris, animaux fantastiques, surtout les éléments déchaînés: tonnerre, éclaire et grêle; *Ibid.*, 309: 'es ist alles Blendwerk, kehre dich nichts dran'.

46. *Ibid.*, 312–13 (HH V).

47. C'est le thème de l'ensorcellement, qui fait disparaître l'objet même de la quête, ou pousse les prospecteurs sur de fausses pistes; sur le diable, maître de l'illusion, Heilfurth, *op. cit.* (12), cite Luther (p. 116); sur les alliés du diable, les Welsches, cf. S.R. Acxtelmeier, *Des aus der Unwissenheits-Finsternus eretteten Natur-Liechts* (Augsbourg, 1699), 86, 'Lorsque l'on arrive aux lieux indiqués et que l'on fouille, on ne trouve rien, car les lieux sont ensorcelés par de mauvaises gens . . . .'.

48. Schneider, *op. cit.* (15), 311-12, HH V, 'Ich binn darinne gewesen, und habe manche Todten-Gebeine daliegen sehen'; parmi les dangers signalés, comme pour toute course de montagne, le mauvais temps et l'incertitude sur les chemins.

49. On peut confronter les légendes nées sur le site (salles souterraines habitées par une faune d'épouvante) résumées in Wilsdorf, *op. cit.* (15), 80-5, avec le 'Walenbuch', où l'explorateur passe devant des figures immobiles, qui l'éclairent ou lui évitent la chute au bon moment, probablement des signes gravés dans la roche: Heilfurth, *op. cit.* (12), n° 987 (cf. Annexes, I).

50. Avant d'entrer dans la zone des signes et repères, il faut bien demander son chemin; ne pas se découvrir pour autant: il faut se méfier des Bohémiens quand on arrive sur la ligne de crête (Schneider, *op. cit.* (15), 304, 'Walenbuch' attribué à Johannes Wale). Dans les parages de Hirschberg, vers 1650, pour ne pas éveiller les soupçons, le prospecteur doit se rendre aux bains de Warmbad et, tout en se baignant, demander des informations sur le village de charbonniers de Seifershau; parvenu dans la Combe de l'Ange, répondre à toute question que l'on cherche des herbes et des racines (*Ibid.*, 309); cf. Heilfurth, *op. cit.* (12), Annexes, III.

51. Schneider, *op. cit.* (15), 306 (HH IV, 'Ich Hans Man . . .'), 'so kommest du auf brüchicht und schwappicht Erdreich, ist auch strittig von grossen Kräutern, und fleusset ein klein Wässerlein verborgen . . .'.

52. Wutke, *op. cit.* (10), 84, 85, 'so wirstu komen yn eyn gestruthe adir weychunge der erdin'; 'doselbist findistu golt slacken und silber slacken manch tausend fudir'. Heilfurth, *op. cit.* (12), n° 991 (Bohême), 'Daselbst hat man im freyen Feld ausgeschmeltzt/Körner und Schlacken liegen in einem tieffen Loch'.

53. Wutke, *op. cit.* (10), 83 ('Ich Anthonius Wale . . .'), si l'on se regarde les uns les autres, on voit des formes bleues en raison des gîtes métallifères et du soufre qui abondent dans le vallon. . . .

54. Wutke, *op. cit.* (10), 84, 'ertreych swartcz', et, sous terre, Heilfurth, *op. cit.* (12), n° 987 (roches noires).

55. Wutke, *op. cit.* (10), 84, 86, 'vorsuche dy steyne mit eyner keylhawe'; 'findistu swartcze gesteyne . . . wenn du sy czuslest, so seyn sy ynnewigk braun alz eyn scharlach . . .'.

56. Schneider, *op. cit.* (15), 309-12 (HH V).

57. *Ibid.*, 308, 'Da wirstu einen veralteten weg sehen . . . er wird sich seltsam drehen'.

58. *Ibid.*, 304, 305, 306, 310.

59. Le 'prototype' de Breslau mentionne l'image d'un évêque gravée sur une roche; le thème du moine revient plusieurs fois dans les 'Walenbücher', mais les seuls signes reproduits dans le texte sont abstraits, symboles de l'or et de l'argent ou marteaux et pointerolles.

60. Heilfurth, *op. cit.* (12), ill. 10, 11 et 12; Wilsdorf, *op. cit.* (15), 275-8. Le doute sur leur utilisation apparaît dans les 'Walenbücher' mêmes, Wutke, *op. cit.* (10), 85 ('ap du en nicht findest . . .'); Schneider, *op. cit.* (15), 305, HH III ('so du es findest') 310, HH V ('die zeichen sind gemacht um der irrung willen').

61. Schneider, *op. cit.* (15), 280.

62. *Ibid.*, 303-4, 'slagksperre', 'da ist ein schacht versetzt; in demselbigen schachte findest du goldstücke . . .'.

63. Heilfurth, *op. cit.* (12), Annexes, I.

64. L'analyse a été conduite par Peuckert, *op. cit.* (16).

65. Les itinéraires sont construits de repère en repère, par un observateur qui effectue des visées lointaines ou proches: distance, couleurs, orientation par rapport au soleil.

66. Schneider, *op. cit.* (15), 291-2 a identifié toutes les grandes roches évoquées

au Sud de Hirschberg; Peuckert, *op. cit.* (16), a trouvé des paillettes d'or en des lieux indiqués par les guides et il décrit le paysage du Spitzstein (233-4).

67. Wutke, *op. cit.* (10), n° 208, 209, 214.

68. *Schlesisches historisches Labyrinth* (Breslau et Leipzig), XXV, cité par Schneider, *op. cit.* (15), 300. Ce dernier auteur oppose dans une formulation frappante 'Erdschätze' et 'Geldwirtschaft' (301).

69. H. Weczerka, *Schlesien Handbuch der historischen Stätten* (Stuttgart, 1977), 506.

70. D. Stobbe, 'Mitteilungen aus Breslauer Signaturbüchern', *Zeitschrift des Vereins für Geschichte und Alterthum Schlesiens*, VI, 350, VII, 358-60, VIII, 439-40. Quant à l'éditeur du texte, K. Wutke, il ne doutait ni de l'identité d'Antonio Wale, ni de son rôle dans l'aventure minière.

71. Aucun ouvrage de synthèse n'a pris le relais de Y. Renouard, *les relations des Papes d'Avignon et des compagnies commerciales et bancaires de 1316 à 1378* (Paris, 1941); pour la Pologne, J. Ptasnik, *Italia mercatoria apud Polonos saeculo XV ineunte* (Rome, 1910); pour la Silésie, H. Wendt, *Schlesien und der Orient, Darstellungen und Quellen zur schlesischen Geschichte, 21* (Breslau, 1916).

72. Wendt, *op. cit.* (71), 46 et suiv.; Ptasnik, *op. cit.* (71), n° 51, 57, 61, 63; D. Stobbe, *op. cit.* (70), VIII, 119 (1424/26/V).

73. Ptasnik, *op. cit.* (71), n° 43.

74. Wutke, *op. cit.* (10), 28, 97; Stobbe, *op. cit.* (70), VIII, 439-40; Ptasnik, *op. cit.* (71), n° 69, 70, 71, 72.

75. Ptasnik, *op. cit.* (71), n° 74, 75, 76.

76. Cf. W. von Stromer, 'Nürnberger-Breslauer Wirtschaftsbeziehungen im Spätmittelalter', *Jahrbuch für fränkische Landesforschung*, 1974/1975, 1079-1100, surtout 1093-4; *Id.*, 'Versuche zur Beherrschung des Weltmarkts für Buntmetalle', *Aspetti della vita economica medievale, Istituto di Storia economica dell' Università di Firenze* (Florence, 1984), 317-41.

77. W. von Stromer, 'Wassersnot und Wasserkünste im Bergbau des Mittelalters und der frühen Neuzeit', *Montanwirtschaft Mitteleuropas vom 12. bis 17. Jahrhundert, Forschungsprobleme*, éds W. Kroker et E. Westermann, *Der Anschnitt, Beiheft 2* (Bochum, 1984), 50-72.

78. Ph. Braunstein, 'Innovations in Mining and Metal Production in Europe in the Late Middle Ages', *The Journal of European Economic History*, 12, 1983, 573-91.

79. F. Graus, 'Die Handelsbeziehungen Böhmens zu Deutschland und Oesterreich im 14. und zu Beginn des 15. Jahrhunderts', *Historica II*, Prague, 1960, 77-110; R. Klier, 'der schlesische und polnische Transithandel durch Böhmen nach Nürnberg in den Jahren 1540 bis 1576', *Mitteilungen des Vereins für die Geschichte der Stadt Nürnberg*, 53, 1965, 195-228.

80. Wutke, *op. cit.* (10), 83: 'Wen ich Anthonius Wale von den gnoden gotis genugk habe an slossirn und an dorffern . . .'; 'do habe ich Anthonius Wale gross gut undir behaldin, daz sich wol mochtin ir hondirt von neren . . .'.

81. Münster, *op. cit.* (11), I, 809. Sur ce point, les remarques de H. Wilsdorf, 'Aspekte der Montanethnographie. Zugleich ein Rückblick auf die Montanarchäologie', *Deutsches Jahrbuch für Volkskunde, Deutsche Akademie der Wissenschaften*, 10, Berlin 1964, 54-71.

82. Rappelons que l'oeuvre capitale de Georg Agricola, *De re metallica libri XII* est parue en 1556, quelques mois après la mort de son auteur.

Contents of Former Volumes

236

A. RUPERT HALL and N.C. RUSSELL, What about the Fulling-Mill?

MICHAEL FORES, *Technik*: Or Mumford Reconsidered.

SEVENTH ANNUAL VOLUME, 1982*

MARJORIE NICE BOYER, Water Mills: A Problem for the Bridges and Boats of Medieval France.

Wm. DAVID COMPTON, Internal-combustion Engines and their Fuel: A Preliminary Exploration of Technological Interplay.

F.T. EVANS, Wood since the Industrial Revolution: A Strategic Retreat?

MICHAEL FORES, Francis Bacon and the Myth of Industrial Science.

D.G. TUCKER, The Purpose and Principles of Research in an Electrical Manufacturing Business of Moderate Size, as Stated by J.A. Crabtree in 1930.

ROMAN MALINOWSKI, Ancient Mortars and Concretes: Aspects of their Durability.

V. FOLEY, W. SOEDEL, J. TURNER and B. WILHOITE, The Origin of Gearing.

EIGHTH ANNUAL VOLUME, 1983*

W. ADDIS, A New Approach to the History of Structural Engineering.

HANS-JOACHIM BRAUN, The National Association of German-American Technologists and Technology Transfer between Germany and the United States, 1884–1930.

W. BERNARD CARLSON, Edison in the Mountains: The Magnetic Ore Separation Venture, 1879–1900.

THOMAS DAY, Samuel Brown: His Influence on the Design of Suspension Bridges.

ROBERT H.J. SELLIN, The Large Roman Water Mill at Barbegal (France).

G. HOLLISTER-SHORT, The Use of Gunpowder in Mining: A Document of 1627.

MIKULÁŠ TEICH, Fermentation Theory and Practice: The Beginnings of Pure Yeast Cultivation and English Brewing, 1883–1913.

GEORGE TIMMONS, Education and Technology in the Industrial Revolution.

NINTH ANNUAL VOLUME, 1984*

P.S. BARDELL, The Origins of Alloy Steels.

MARJORIE NICE BOYER, A Fourteenth-Century Pile Driver: the *Engin* of the Bridge at Orleans.

FOURTEENTH ANNUAL VOLUME, 1992

*Out of print